Personal Communication Systems and Technologies

The Artech House Mobile Communications Series

John Walker, *Series Editor*

Advanced Technology for Road Transport: IVHS and ATT, Ian Catling, editor

An Introduction to GSM, Siegmund H. Redl, Matthias K. Weber, Malcolm W. Oliphant

Cellular Radio: Analog and Digital Systems, Asha Mehrotra

Cellular Radio Systems, D. M. Balston, R. C. V. Marcario, editors

Cellular Radio Performance Engineering, Asha Mehrotra

Mobile Communications in the U.S. and Europe: Regulation, Technology, and Markets, Michael Paetsch

Land-Mobile Radio System Engineering, Garry C. Hess

Mobile Antenna Systems Handbook, K. Fujimoto, J. R. James

Mobile Data Communications Systems, Peter Wong, David Britland

Mobile Information Systems, John Walker, editor

Narrowband Land-Mobile Radio Networks, Jean-Paul Linnartz

Personal Communications Networks, Alan Hadden

Smart Highways, Smart Cars, Richard Whelan

Wireless Data Networking, Nathan J. Muller

Wireless: The Revolution in Personal Telecommunications, Ira Brodsky

For a complete listing of *The Artech House Telecommunications Library,* turn to the back of this book.

Personal Communication Systems and Technologies

John Gardiner and Barry West
Editors

Artech House
Boston • London

Library of Congress Cataloging-in-Publication Data
Personal communication systems and technologies / John Gardiner and Barry West
Includes bibliographical references and index.
ISBN 0-89006-588-8
1. Mobile communication systems. 2. Wireless communication systems. 3. Celllular radio.
    4. Local area networks (Computer networks). I. Gardiner, John. II. West, Barry.
TK6570.M6P48 1995   94-23867
384.5–dc20   CIP

British Library Cataloguing in Publication Data
Personal Communication Systems and Technologies
I. Gardiner, John II. West, Barry
621.38456

ISBN 0-89006-588-8

© 1995 ARTECH HOUSE, INC.
685 Canton Street
Norwood, MA 02062

International Standard Book Number: 0-89006-588-8
Library of Congress Catalog Card Number: 94-23867

10 9 8 7 6 5 4 3 2

Contents

Foreword

By the 1960s, there were tens of millions of mobile radio units in the service of commerce, transportation, and security. But mobile radio remained relatively isolated from public telecommunication networks and mass demand until the advent of public cellular systems in the mid-1970s. Cellular systems, along with contemporary advances in information technology and microwave techniques and the important evolution of regulatory policies, unveiled a great need and market for personal communication. But first-generation cellular service also exposed the limitations of the congested available spectrum and 25-kHz-bandwidth FM technology to supply adequate voice quality, capacity, coverage, and connection reliability. Cordless telephones connected to home or PBX terminals further stimulated public demand for mobility.

Second-generation cellular systems, which benefit from expanded frequency allocations and digital technology, show improved performance, increased traffic capacity, and a greater scope of services, thus easing the way to personal communication services (PCSs). Financial pages of daily journals confirm the economic importance of these systems to users and investors. GSM is an important realization of second-generation, portable communication for Europe and many other regions. Different digital implementations are seen in North America (where both TDMA (D-AMPS) and North American CDMA are used) and in Japan (JDC). Globally, all four of these systems are recommended by the International Telecommunication Union (ITU). These systems are not compatible and do not permit "roaming" outside of the area of use of a particular standard. The acceptance of digital systems is growing and further accelerating the expansion of mobile markets. A number of the earliest PCS services are being implemented on the basis of these standards.

Advanced mobile service suppliers look to a third-generation of cellular systems to offer voice quality that is fully comparable to the highest standards of public networks; an appropriate range of video and data service; comparable performance in portable, vehicular, rural, urban, and indoor situations; and provisions for wider "roaming" (possibly global). Experts and organizations from all regions participate in the ITU's studies and its Recommendations on Future Public Land Mobile Telecommunications Systems, provisionally entitled *International Mobile Telecommunications-2000 (IMT-2000)*. (See Chapter 12.) Integrated, complementary use of satellite and terrestrial technologies is foreseen, with full exploitation of intelligent network capabilities. Recommendations have already been agreed upon for services to be supported, network architectures, satellite operation requirements for the radio interface and functionality, and spectrum considerations. One result of these recommendations was the allocation by the World Radiocommunication Conference of 1992 of new worldwide frequency bands for mobile service that provide 230 MHz of additional spectrum near 2 GHz. Actual use of these bands may vary in different countries, but they are available globally. Studies continue on radio access, spectrum and network considerations, the regulatory environment, and the relationship with the network universal personal telecommunications (UPT) concept.

The authors provide both a technical perspective on personal communication systems and an overview of the market for these systems. While Chapters 4 through 6, which discuss real and emerging systems, deal largely with European systems, these systems are being introduced in many other countries. Chapters 7 through 13 are written from the North American and Asian perspective; Chapter 12 is a timely overview of the progress toward regional and global standards.

In radio's first century, there was little need or spirit for the standardization of different, independent mobile radio systems. The experience gained during a few short years of network-oriented mobile communications brings the questions of compatibility for global personal communications into sharp focus.

Richard C. Kirby

Preface

In preparing a text on any developing technology, one is faced with the need to "stop the clock" and encapsulate the state of the art at that instant. The huge potential of personal communications as a global industry for service providers, network operators, and equipment manufacturers has attracted vast resources in research and development in Europe, North America and the Far East, and the resulting rate of progress is impressive. Our aim in this book is to present a view of the evolutionary process which has produced the personal communications networks currently being implemented, and which will deliver the next, "third" generation of systems.

Part I, comprising Chapters 1–3, sets the scene by examining the role of personal communications from the standpoints of social impact and expected demand; the regulatory background to the pioneering initiative taken in the UK, which paved the way for other European systems; and the constraints imposed by the characteristics of the radio environment.

In Part II, we first examine in Chapters 4–6 the extent to which second-generation systems—GSM, DCS1800, and DECT—are able to support the personal communications requirement. Next, in Chapters 7–9, the discussion is extended to consider other modes of service delivery which will help to provide the desired continuum of service capability to adaptive multimode user terminals, as the user moves between different environments.

Part III, comprising Chapters 10–12, presents perspectives of present and future developments in the US, Japan, and Asia, and concludes by surveying the current debate regarding the way forward to global standards and truly international roaming capability.

The history of systems supporting radio-connected personal terminals has been brief in comparison with other sectors of the telecommunications industry. Undoubtedly the next decade will see some of that history rewritten, but it is hoped that this text will equip the reader to comprehend new developments, based on an understanding of the current state of the art and a perception of the direction of its progress.

The editors wish to record their thanks to the authors for their hard work, and to the series editor, Dr. John Walker, for his unfailing enthusiasm and support.

John Gardiner
Barry West
January 1995

Part I
Background and Fundamentals

Chapter 1

The Needs and Expectations of the Customer

John Gardiner and Barry West

Writers of fiction [1,2] have sometimes depicted future societies in which technology, especially telecommunication technology, has come to have a sinister, rather than a wholly beneficial, significance in people's lives.

Such a view of an intrusive and pervasive information environment, in which individuals are dominated by the technology they themselves have created, makes the concept of personal communications appear more a threat than a promise. However, the transition that is now taking place as personal communications evolves from a specialized minority service to a generally available one, merely marks a natural step in the process of maturity in telecommunications. Of course, society and the individual will want to maintain control over how this technology is used so as to reap its benefits without suffering any negative impact on our lives.

It has always been obvious that telecommunication should preferably be able to follow the user around, but only lately has the required technology (or in truth, the combination of several technologies) developed to the stage where this science-fiction idea can become fact.

This book gathers together views of where personal communications is now and where it is going. In this first chapter, we survey the issues—technical and other—arising in the implementation of these systems.

1.1 PERSONAL COMMUNICATIONS: A SERVICE OR A SYSTEM?

It is appropriate to consider at the outset what will be the needs and expectations of the customers for personal communications. The requirement for personal communication services arises from the changing environment at the end of the 20th century, which is bringing changes in life and work styles in the developed world; we should also not forget the burgeoning recognition of basic communication needs of the third world. Because the emerging technology addresses real human needs, we can be confident that personal communications will not be just another stillborn technology or sterile market prospect.

The main force for change, however, does stem initially from the business world, and several reasons for this can be cited:

- Changing employment structures and increased mobility of labor are leading to the decline of occupationally centered communities and a growth in dispersed working centers with common interests, linked by customized communications and often sharing common databases.
- Rising transport and accommodation costs encourage the relocation of industry and more people working from their homes. Those with specialist skills will assume more flexible patterns of employment.
- Demand for more—and more effective—communications will grow as organizations and individuals respond to increasing social and market pressures.
- Greater pressure for training and retraining to bridge skill gaps.
- Increase in telecommunicating as an addition to, rather than a replacement for, meeting at a central place.

What, then, are the implications of this perception of the future? Taking the existing situation in the United Kingdom as a starting point, a typical customer might wish to access voice, fax, and data services while on the move, either as a pedestrian or in a car, train, or plane. In this case, the required services can be obtained from a multiplicity of physical networks and service providers. In an office environment, a wireless private exchange may provide connection to one of a small but growing number of fixed public networks; the choice of network will be determined by the local subscription arrangements. There may also be a standalone wireless local area network available for computer workstation interconnection. Out-of-doors, two cellular radio operators and, now, two personal communications network (PCN) operators can offer voice telephony with additional services (depending on the choice of provider) such as fax and voice mailbox; again, the customer will choose to subscribe to one of the networks. Alternatively, a "telepoint" service could offer similar basic facilities in localized areas around public access points, but via yet another service provider. As an airline passenger, the customer will be offered a cordless handset for voice telephony which transponds the cordless environment within the aircraft cabin via satellite or terrestrial base stations to the international terrestrial fixed networks.

The shortcomings of this situation are very clear in that the customer is only too aware of separate elements of the technology that deliver the required services and the need for user interaction with many different entities to access them.

The closing years of the 20th century will see a society where communications is increasingly seen as a facilitator in a complex world and as a source of competitive advantage. The current distinction between telephony and computing will disappear as communication becomes the integrated transmission of information in several media (voice, data, image, and text); customized personal communicators developed from the concepts of personal digital assistants (PDAs) will offer that choice of media. This vision does not differentiate between fixed and mobile personal communications, since customers will expect their service needs to be met by both. Likewise, the distinction between public and private communications will be blurred as mobile calls transit from the street to the office with, conceivably, an in-call handover between the corresponding public and private networks.

The transition from the current situation to the future one envisioned here represents a major step forward. The obvious question is whether the advances demanded by the business community point toward sufficient general take-up to warrant the enormous inherent costs. The history and the projected growth of mobile communications are strong indicators that there is massive potential in the personal communications industry. Whereas in the developed world, telephone penetration is typically 50% of population, the market for radio-based personal terminals could ultimately approach one per adult, say, 80% of the population.

Taking western Europe as an example, at the end of 1993 there were about 8 million users of cellular telephony, and this figure is increasing by 30% to 40% per year. There are about the same number of users of other radiocommunication systems, including paging and private mobile radio (PMR) systems, although this market sector is growing more slowly. Forecasts of the growth of the total market vary but indicate that the number of users of mobile radiocommunication could exceed 40 million by the year 2000. Thus, the indications are of a considerable unsatisfied appetite for telecommunication of the sort that personal communication systems can provide.

1.2 THE TECHNICAL CHALLENGE

1.2.1 Recognition of the System Problem

If personal communications is essentially a service concept, then it is obvious that fundamental changes are necessary in the ways in which the physical components of existing networks are made to interact and interoperate. The future personal communications "system" will have to provide an integrated technical environment through which service providers can deliver their products to the users, who will access whatever they require from a single terminal, migrating (roaming) between providers and using whatever combination of transmission and reception parameters are appropriate. This vision has been widely accepted in many different technical communities and the international regulatory community. Its importance for the future of the European telecommunications industry has been spelled out most recently in the European Commission Green Paper [3], "Towards the Personal Communications Environment," which concludes that to promote evolution toward personal communications services, ". . . the basic requirements are to remove, initially, restrictions on the combination of multiple mobile technologies or services through a single service provider, and subsequently, to remove restrictions on the free combination of services provided via fixed and mobile networks.''

Within the technical community, the problem has been approached from several different standpoints in parallel, and the situation has been further complicated by the ways in which research and development work has been supported by diverse mechanisms in Europe and in wider international groupings. The response of the fixed network community has been to progress toward universal personal telecommunications (UPT). This means that the fixed networks will acquire the functionality currently associated with the public land mobile network (PLMN) components of cellular radio systems, that is, the ability to register the location of an individual subscriber regardless of the point of access to the network. On the radio side, interest has increasingly turned to the need to make services available to users in several different radio environments that require differently optimized air-interface parameter sets, but accessed by a single terminal.

1.2.2 The Generations of Personal Communications

In historical terms, the progress from first-generation to second-generation mobile communications systems and the ongoing process of defining third-generation systems can be seen as follows. First-generation systems were the first to enable a mobile user to access and use continuously a telephone connection anywhere in the service area of a mobile network operator. This continuous, extensive coverage required a ''handover'' (or ''handoff'') mechanism between individual cellular coverage areas from each base station (cell site). Such systems include the advanced mobile phone system (AMPS) in the United States; the total access communication system (TACS), a modified version of AMPS used in the United Kingdom and elsewhere; and the Nordic mobile telephone (NMT) system, prevalent in the countries of northern Europe. While not having the handoff function, the first generation of cordless telephone equipment, for example, cordless telephone 1 (CT1), can be included since it was an important progenitor of later developments by virtue of the utilization of a small lightweight handset. The first-generation systems were characterized by having the voice signal transmitted on the radio carrier in analog form and by starting as national standards without cross-border, or sometimes even extensive national, coverage. (An exception was NMT, which from the outset targeted coverage of the Nordic countries.)

Second-generation systems arose from the success of the first-generation ones and built on some of the lessons learned in the technical, commercial, political, and regulatory fields.

Technically, there was the need to improve the efficient use of the radio frequency spectrum (more users per megahertz), coupled with a desire to use digital encoding of the voice and to use digital modulation of the radio bearer. Politically, a number of nations saw mobile telecommunications as a trial, or the leading edge, of an increasing desire to move away from a monopoly supplier of telecommunications. This led in turn to the use of regulation (in a basically deregulated situation) to generate and stimulate competition. The competitive element and the need for profitability ensured that the commercial control took account of perceived user needs, with a growing change from technology-led development to those led by customer requirements.

The lead in second-generation systems was taken in Europe by the European post and telecommunication operators' organization (CEPT), which created the Groupe Spécial Mobile (GSM Committee), later transferred to the control of the European Telecommunications Standards Institute (ETSI). The abbreviation GSM came to be used as the name of the new systems and was adopted by European operators to stand for "global system for mobile (communications)." ETSI, whose official language is English, later changed the name of the committee to Special Mobile Group (SMG). GSM was given the task of devising a mobile telecommunication system that would operate to the same standard throughout Europe and allow users to access and use the system anywhere irrespective of home region or equipment supplier.

The major distinctive feature of second-generation systems, of which GSM is the foremost example, is the use of all-digital techniques in the air-interface. This strategy, discussed in more detail in Chapter 2, represented a watershed in the evolution of personal communications and provided a clear separation between the first-generation analog world and the second-generation digital one.

If GSM was the second-generation *cellular* system, then what of other second-generation developments? In Europe, the highly proactive role assumed by the European Commission and its interaction with CEPT and ETSI, resulted in powerful stimuli toward definition of other parallel second-generation systems. Cordless telecommunications saw the appearance of CT2 technology in the United Kingdom initially, and cordlessness was addressed next as the digital European cordless telecommunications (DECT) standard began to take shape. In the PMR arena, the trans-European trunked radio (TETRA) system commenced its evolutionary path through ETSI, while the paging industry looked toward a European standard in the European radio messaging service (ERMES). A further emerging user need and potential market was identified in providing service to airline passengers via terrestrial base stations, and the terrestrial flight telephone service (TFTS) started its progress to standardization. The second-generation picture was ultimately completed by the digital short range radio (DSRR) system, which caters to the needs of users whose communication requirements do not include connection to the fixed network.

If the transition from analog to digital technology marked the progression from first- to second-generation systems, what will distinguish third-generation personal communications? The answer to this question has two principal facets, which are, broadly, "technical" and "political." The technical issues are those already mentioned, which will enable users to migrate freely among the many service offerings and supporting networks. The political aspects are discussed in more detail in Chapter 12 and relate to the international impact of third-generation systems. In embarking on the definition of GSM and other second-generation systems, a major step was taken in bringing together all the interested administrations in CEPT to arrive at a consensus on the technical standards for the systems and, with the aid of the European Commission, on the spectrum to be reserved throughout the CEPT community for their introduction.

Extrapolation of this approach would target the evolution of the third-generation personal communications environment as a regional standard satisfying the needs of Europe as represented by the now enlarged CEPT community of 31 European administrations. As thinking on the post-second-generation world began to consolidate in Europe in the mid-1980s, the European Commission compiled a Green Paper [4] and a follow-up report [5] that assumed this regional focus and initiated the pan-European research program R&D in Advanced Communications Technologies in Europe (RACE), which would lay the foundations for the third-generation environment in Europe. Implementation of the results of RACE research would establish an integrated broadband communications network (IBCN) as a successor to the integrated services digital network (ISDN), and this would support large numbers of radio-connected terminals. The mobile community would operate in an environment known as the universal mobile telecommunications system (UMTS).

However, as RACE began to get under way, a parallel program (described in more detail in Chapter 12) was initiated in the (then) International Consultative Committee for Radio (CCIR), which set about considering a possible global standard for mobile and personal communications. The initial perception was of a framework for the truly international system generations of the longer-term future. This program aimed to define a "future public land mobile telecommunications system" (FPLMTS); it has recently been proposed that this term be replaced by the less awkward International Mobile Telecommunications-2000 (IMT-2000).

Further parallel activity had also been under way in North America and Japan to define their versions of third-generation systems, and these other developments (the subjects of Chapters 10 and 11) also formed a backdrop to the work on global standards within the International Telecommunication Union (ITU) community, centered in Task Group 8-1 of the ITU Radiocommunication Sector (ITU-R).

Inevitably, in such a volatile technology, perceptions and emphasis are continually being revised in the light of events, and the rather simplistic initial view of a clearly defined third-generation UMTS followed by a further completely new fourth-generation global standard is now open to question. Since major elements of the GSM standard were finalized and equipment started to appear, more than 60 countries worldwide have decided to adopt the system with the result that GSM is becoming a de facto global standard. Moreover, the potential for evolution of second-generation systems to deliver the capabilities originally conceived for UMTS has made it increasingly difficult to define clear boundaries between the second and third generations. Some of the difficulties associated with this situation are best appreciated by reference to the kinds of development now taking place.

1.3 TECHNICAL DEVELOPMENTS

1.3.1 Directions in Radio Technology

Recalling the problems encountered by our customer in Section 1.1, an obvious question is, "Why have second-generation systems emerged with air-interfaces that differ so widely in their physical layer parameters?'' The next question then would be, ''How can a third-generation air-interface be defined that would be satisfactory for all applications?'' Part II of this book sets out the reasons for the current situation by showing how different combinations of modulation, speech and channel coding, multiple access, and duplex strategy are optimum in different environments and service applications. Considering only digital systems, it is easy to see that, for instance, in a short-range cordless application, short radio propagation paths are likely to result in delay spreads that are too small to justify channel equalization. Similarly, time-division duplex on a single radio bearer is feasible for DECT and CT2, since the guard times needed to allow for propagation delays are insignificant. By contrast, GSM, in catering to large cell requirements, must use channel equalization to combat intersymbol interference from widely differing multipath delays and needs to operate in two-frequency duplex to permit separate timing adjustment on the mobile up-link.

Other considerations, such as the differing digital error-rate requirements set by different services, further complicate the problems of defining a single set of air-interface parameters to satisfy all user requirements.

New technical possibilities are also being introduced into the optimization process. Of particular topical interest at the time of this writing is the vigorous debate surrounding the choice of multiple-access strategy. When basic system choices were being made for GSM, the viability of spread spectrum techniques were questionable, given the anticipated status of the semiconductor technology in delivering low-power-consumption chips with high levels of processing capability in the required time scale. Code division multiple access (CDMA) techniques are now seen to be feasible and attractive in the context of third-generation systems, which may be standardized at the end of this decade. Equally, the choice of modulation scheme for GSM (Gaussian prefiltered minimum shift keying, or GMSK) was dictated by the need for a constant envelope modulation that could be amplified without spectrum dispersion in power-efficient but nonlinear power amplifiers. Developments in amplifier linearization techniques have relieved this constraint, and more spectrally efficient but nonconstant envelope modulations are now viable contenders for third-generation systems. Pi/4 OKQPSK and 4 or 16 QAM are under serious consideration for situations in which the quality of the radio channel will permit their use with acceptable error rates.

Overall, the emerging picture is of a range of diverse radio solutions, each optimized for a combination of service requirements and operating conditions. The need then is for user terminals to acquire the capability to adapt their operating mode according to the services required at any one time and the available radio environment.

1.3.2 Developments in Fixed Networks and Services

Reference was made in Section 1.2.1 to the aspect of fixed network evolution now known as UPT. But this is only one of a number of major advances in fixed network technology that will have substantial impact on the way in which third-generation systems emerge.

Progress continues to be made in wide-band transmission capability with the exploitation of increasingly powerful optical technology, and this brings the promise of proportionally increased bandwidth at subscriber terminals. These developments point strongly toward the need for those concerned with the radio technology of future systems to anticipate the demands of users who will have become accustomed to having access to very high bit-rate services in hard-wired terminal environments and who will expect similar facilities to be available to them via their personal communicators.

Aside from basic bit-rate considerations, the UPT capability is one manifestation of the evolving *intelligent network* concept emerging from research and development in the management of databases and in the process of service creation. Returning to the open network access perception of the future as set out in the 1993 Commission Green Paper, it is evident that progress has to be made toward separating the fixed network infrastructure that delivers services to users from the activities of the service providers. The latter will create services to meet the needs of users and will market and deliver them by buying access to the transmission infrastructure.

Given the increasing bandwidth capability of the network, other aspects of the transmission process are becoming progressively more significant. The introduction of asynchronous transfer mode (ATM) and the general trend toward packet-switched, or connectionless, transmission continues to present new challenges to the radio system community, which must devise ways of transporting ATM-based packet switched services into its air-interfaces. The added features of synchronous digital hierarchy (SDH) and frame relay as transport mechanisms, which are also becoming established in the fixed-network environment, all point toward the need for research and development activity aimed at third-generation mobile systems to be integrated with that relating to fixed networks.

1.3.3 Developments in Information Technology

A criticism of the foregoing arguments is sometimes voiced by sections of the current user community who have difficulty perceiving the uses for the new high-capacity telecommunications environment. These reservations usually derive from a bottom-up approach of mapping existing services onto a supposed future environment, which is not an appropriate way of looking at the evolutionary trends. Third-generation systems will have lifetimes that will extend well into the second decade of the next century. Users in the 21st century will take for granted access to information on a far greater scale than that envisaged by today's users. The potential of multimedia services is just beginning to make an impact on the fixed-network user community as the process of converging the communications and computing environments begins to gather momentum. Extrapolation of this process leads to a perception of the future personal communicator as a mobile office suite, a personal organizer, and a personal manipulator of the information that will be essential to the efficient operation of both business and social interactions. The term *personal digital assistant* is already current and will assume progressively greater significance as time goes by.

1.4 PROGRESS TOWARD UMTS AND FPLMTS/IMT-2000

1.4.1 The RACE Program

RACE was a research initiative aimed at the development of telecommunication systems for Europe. It was conceived in 1985 as part of the European Commission Second Framework Program, whose stated objective is "to prepare for the introduction of Integrated Broadband Communication . . . progressing to community wide services by 1995."

The history of the RACE program offers an interesting insight into the developing significance of mobile and personal communications since the late 1980s. In the first phase of RACE research, which began in 1987, the emphasis was on the development of the fixed broadband network, the IBCN. A single project (R-1043) was included to consider the implications of (and for) mobile communication systems in relation to those fixed networks. It was in fact this project that first developed the idea of the UMTS, which underlies much of the more recent thinking about mobile systems.

By the time RACE II was launched in 1991, the significance of mobile communications had increased to the extent that it became one of the central themes of the program, and the following eight projects were included dealing with different aspects of mobile systems.

- PLATON (planning tools for third-generation networks) has the objective of developing software tools that will allow the spectrum efficient planning of UMTS networks.
- MONET (mobile network) aims to define a fixed infrastructure for UMTS as part of IBCN, permitting maximum exploitation of the intelligent network and commonalities with UPT and ensuring that a wide range of mobile services can be supported in an economic way.
- CODIT (code division testbed) is exploring the potential of CDMA for UMTS, addressing such issues as radio interface parameters, macrodiversity, fast and soft handover, microcells and picocells, and radio network planning methods.
- ATDMA (advanced TDMA mobile access) is concerned with identifying optimum ways of interfacing narrowband mobile services with the IBCN and with addressing the issue of developing an adaptive time-division multiple access (TDMA) air-interface to accommodate different service requirements and radio environments.
- MAVT (mobile audiovisual terminal) seeks to develop a powerful video coding algorithm for transmission of moving and still video in a mobile environment and to implement it, together with an audio coding algorithm, on a demonstrator.
- MBS (mobile broadband system) anticipates the need for broadband services in the third-generation environment and is concerned with defining these services and

matching user needs to cost-effective technology. Since the 60-GHz region of the spectrum appears to be best able to provide the necessary bandwidth, subsystem technology is the principal focus.

- SAINT is concerned with the integration of satellite and terrestrial mobile/personal communications networks.
- TSUNAMI is researching the role and the impact of adaptively steerable antennas in future system architectures.

Two trends are clear. First, the recognition that the needs of the radio-connected community of users are now a major factor in the evolution of third-generation systems. Second, overall system integration issues and the development of the third-generation system as an integrated whole have been recognized as having foremost significance.

1.4.2 The European Fourth Framework Program

The RACE program has been one of many pan-European research efforts supported under the overall umbrella of the Third Framework Program, which is now drawing to a close. Planning for a Fourth Framework Program is now well advanced; within it, a new program, Advanced Communications Technologies and Services (ACTS), will carry forward the research completed in the RACE program. The overall strategy for ACTS continues the theme of integrated broadband telecommunications in Europe with a large proportion of radio-connected users drawing on a greatly expanded range of services. Six areas of work have been defined:

- Interactive digital multimedia systems and services;
- Technologies for photonic networks;
- High-performance networking;
- Mobility and personal communications networks;
- Intelligence in networks and service engineering;
- Quality of service, security, and safety of communications services and systems.

1.5 SOCIAL AND ENVIRONMENTAL CONSIDERATIONS

The introduction of any innovative technology that is likely to be experienced by a significant proportion of the population is liable to raise concerns about potential adverse effects on the environment and the quality of people's lives. The more inaccessible or "revolutionary" the technology, the greater these concerns tend to be. In the case of personal communications, public disquiet focuses on the following issues:

1.5.1 Visual Impact

There are natural fears that base station installations will proliferate and have unsightly masts, antennas, and equipment cabins. In rural areas, where the economics of service provision tends to favor large cells and relatively elevated antennas, much can be done to mitigate the visual impact by careful siting in relation to natural or existing features (trees, buildings). In suburban and urban environments, however, progress toward higher-capacity systems means progress toward smaller and smaller cells and base station installations that will be no more obtrusive than wall-mounted security alarm equipment or other "street furniture."

1.5.2 Security, Privacy, and Intrusion

Security is a broad term that covers the control of access to information, resources, or services in a network. It includes means to prevent fraudulent use, for example, the impersonation of a valid user. *Privacy* refers more specifically to access to subscriber information, including identity and location, as well as interception of the call itself. In general, the effectiveness of security features is determined by a combination of system design, implementation, and regulatory control. Sometimes, the protection of privacy may conflict with other requirements. For example, government agencies, including the police, may have reasons for wanting to intercept messages or to discover the location of a person, which may be possible only if limits are placed on the security of a telecommunication system.

Intrusion is a much vaguer term and refers to telecommunication that is undesired, perhaps because it is at an inconvenient time or the content is unwelcome (i.e., junk mail). If people are to accept an increasing amount of telecommunication, they will need ways of preventing it from taking over their lives. This may mean, for example, greater use of "off-line" modes, such as text and voice messaging.

1.5.3 Health and Safety Considerations

For many people, *radiation* is an emotive term regardless of what kind of radiation is at issue. Radio waves are classified as nonionizing radiation and are in principle much less liable to cause harmful effects than the ionizing radiation from radioactive materials, which is sufficiently energetic to disrupt material at the atomic level. Even so, it is known that exposure to very high power electromagnetic fields, such as those found close to radar antennas, can be harmful by causing localized heating. There is, however, no evidence that the much lower powers typically used in radiocommunication systems can cause harm. Standards are established nationally and internationally that limit the allowable exposure of people to such radiation and all properly designed systems operate well within these limits. The standards are in general based on the current scientific understanding of physiological implications of these heating effects.

Some concern remains, however, that there might be effects that occur at even lower levels than the standards allow, perhaps through some (as yet unknown) athermal electrochemical effect in biological cells. Of course, even if there are such effects, it is by no means certain that they would be harmful. Considerable research is under way in this area, supported by national and international funding agencies, but at the time of writing, no definite scientific evidence is known to the authors that any such mechanisms exist.

Clearly, however, as a matter of good practice, prudence requires that the power emitted by radio transmitters of all kinds be at the minimum level consistent with proper operation.

1.5.4 Electromagnetic Interference and Compatibility (EMC)

As electronic equipment becomes increasingly part of everyday life for most people, care must be taken to ensure that equipment designed to radiate electromagnetic energy does not interfere with the operation of other electronic equipment. Airline passengers are now accustomed to being told that use of portable telephones is prohibited onboard because of the possibility that they might interfere with the aircraft's electronics. In Europe, a Commission directive is in force that specifies levels of robustness against interference, but it is to be expected that EMC considerations will figure in the design of third-generation systems and, indeed, may influence choices of radio parameters for UMTS.

It would be wrong to consider these negative aspects in isolation however. Of course, everything should be done to minimize the environmental impact and the probability of harm. But even if some risks remain, they should be set against the considerable environmental and other benefits that personal communication offers. For example, a car driver may occasionally be distracted by his radiotelephone and so cause an accident, but mobile or portable phones are frequently used to summon emergency services to accidents, and lives and property are saved as a result. Furthermore, telecommunication can itself be a substitute for more dangerous (and environmentally damaging) activities, such as physical travel.

A further benefit is the potential value of personal communication systems to people with special needs, including those with impaired hearing, speech, sight, or physical mobility. The existence of a mass market should make it feasible to take such specialized requirements into account in the implementation of systems and services. To achieve this, these needs must be included in the definition of services at an early stage in standardization and system specification.

For this reason, specific tasks to identify and define such requirements are likely to be included in the European ACTS research program.

1.6 SUMMARY

The purpose of this chapter has been to put into context the detailed and specific material in the rest of the book. In the remaining chapters of Part I, we consider the environment in which personal communication systems are evolving. Chapter 2 describes the commercial and political backgrounds to the inception of PCN networks in the United Kingdom and highlights some of the technical issues that bear on the provision of such a service. Chapter 3 reviews the fundamentals of radio propagation and the resulting constraints on system implementation.

Part II is devoted to different aspects of second-generation systems with some forward look toward the application of spread spectrum techniques and the role of satellites in the fully integrated form of UMTS.

In Part III, current developments in United States and Japan are reviewed to put into perspective the possibilities that form the background of the work of arriving at standards for third-generation systems, whether they have a regional focus, as originally conceived for UMTS, or a global one, as targeted by the ITU TG/8-1 community. The final chapter sets out the current position in the standardization process.

REFERENCES

[1] Orwell, G., *Nineteen Eighty-Four*, London: Secker and Warburg, 1949.
[2] Pohl, F., ''The Age of the Pussyfoot,'' *Galaxy Magazine*, 1965. Longer version in *Bipohl*, New York: Ballantine Books, 1982.
[3] Commission of the European Communities, Directorate General XIII, ''Towards the Personal Communications Environment: Green Paper on a Common Approach in the Field of Mobile and Personal Communications in the European Union,'' Dec. 1993.
[4] Commission of the European Communities, ''Towards a Dynamic European Economy: On the Development of the Common Market for Telecommunication Services and Equipment'' (Green Paper), June 1987.
[5] Commission of the European Communities, ''Towards a Competitive Community-Wide Telecommunications Market in 1992,'' Feb. 1988.

Chapter 2

Personal Communication Networks: Concepts and History

John Gardiner and Stephen Temple

Chapter 1 discussed perceptions of personal communications services from the standpoint of current thinking on the role and relevance of personal communication systems (PCS). To put the origins of the U.K. PCN initiative in perspective, it is necessary to recall the atmosphere and aspirations of the telecommunications community at the beginning of 1989. Two influences were at work: the growing conviction that technical possibilities previously regarded as excessively complex could now be realizable in terminal equipment costing no more than a video recorder and the consequences of these perceptions in carrying through innovations in competitive service provision to stimulate further reductions in costs to consumers and extend the basic principles of competition in the provision of telecommunications services.

2.1 THE NEED FOR NEW SPECTRUM

The term *personal communications networks* (PCNs) was coined by the U.K. Department of Trade and Industry (DTI) in the consultation paper "Phones on the Move" published in January 1989. The discussion document came about as a result of the enormous success of the two U.K. cellular radio networks run by Vodafone and Cellnet. Growth in the subscriber base was impressive. While prices were relatively low compared with many other countries, there was a strong case for new entrants to come into the market to create a downward competitive pressure on prices. With advice from the telecommunications watchdog body OFTEL, DTI decided that the U.K. market needed more competition. There was no shortage of commercial undertakings wanting licenses to run public mobile radio networks of the cellular sort. DTI's Radiocommunications Agency carried out a search for radio frequency channels to accommodate a third cellular radio operator. The result of this study was to identify two options. The first was to try to squeeze an extra operator into the 900-MHz part of the radio spectrum near the two existing cellular radio operators. The spectrum would have fallen in the range planned for the new GSM system, which had already been allocated to the two existing U.K. operators. The second option was to look much higher in the radio spectrum and break open a new range of frequencies. Considerably more spectrum was identified as available around 1.8 GHz. In fact, sufficient spectrum was found for two or even three new operators.

The 900-MHz part of the spectrum would have had the advantage of propagation characteristics as favorable as the two existing operators enjoyed, and mass-produced components were readily available. But spectrum on a national basis could not be found without overriding the commitments already given to Cellnet and Vodafone on their use of the GSM bands. Even if it had been possible, there remained other considerations. For example, to have had three rapidly expanding operators squeezed into a localized part of the spectrum would have led, at some point in the future, to spectrum shortage. Indeed, the whole history of mobile radio had been one of a shortage of spectrum limiting growth and keeping prices up. To reverse the trend, there had to be enough spectrum to allow a very large subscriber base, which would create economies of scale and lead to lower prices and hence to an even larger subscriber base. And so, in principle, this virtuous circle would create a consumer mass market. Thus, there was an inevitability that new spectrum would have to be found sooner or later to fuel the growth of cellular radio into the next century. The question was when. Or, more particular to the decision facing DTI in 1988, was it premature to try to open up a new spectrum region at that time?

The disadvantages were obvious. The propagation characteristics at 1.8 GHz would be less favorable than at 900 MHz. Although there were no intrinsic technical problems with 1.8-GHz mass-produced civil mobile radio components, there just did not happen to be any around. This led in turn to consideration of whether systems could be conceived that were not only technically and economically viable but that would lead to competitive pressure on the two existing cellular radio operators.

DTI reviewed a number of technical trends that appeared favorable. The most significant was the trend toward handportable mobiles in the two existing cellular radio networks. Cellular radio had begun life as mobile radios in cars. In fact, *carphone* was a popular term used at the time to describe a cellular radio. Within two years or so of the start of the U.K. TACS cellular radio networks, handportable radios had reached 20% of new connected subscribers. Projections of the trends showed that in the 1990s handportables were set to become the dominant component of new connected subscribers. This prediction proved correct—by 1992, 70% of new terminals sold by one of the cellular operators was a handportable.

The second significant trend was toward small cells. At the outset, the cellular operators were trying to provide national coverage as quickly as possible. This was achieved with large cells that depended on the high power capabilities of vehicle-mounted sets. Handportable coverage was limited to city centers. However, the large buildup of subscribers soon had the two cellular operators splitting cells to cope with the traffic increase. Every cell split reduced the average cell radius and enhanced proportionately the viability of handportable radios. Over the typical radius of small cells in city centers, the propagation difference between 900 MHz and 1.8 GHz would be much less pronounced. What was a possible competitive liability with 1.8 GHz could be neutralized, at least for urban coverage. This still left the more rural areas to be covered and was recognized in the allowance given by DTI for infrastructure sharing between the competitive PCN operators.

A third trend was the advances in very large-scale integrated circuit (VLSI) technology that made handportable radios more and more sophisticated. The profusion of miniature electronic calculators, diaries, dictating machines, and other such consumer items pointed to a future of what the DTI "Phones on the Move" document referred to as the "office in the pocket." The smaller pocket telephones in the TACS market were part of the same trend. Once the cellular-radio chip sets became commodity items, integration of a multiplicity of functions seemed a likely way in which the industry would differentiate their products in the marketplace. Such pocket telephones with various add-ons could become as commonplace as pocket calculators and electronic diaries. Hence, the term *personal communications* seemed to best capture where the market was likely to be by the mid- to late 1990s.

These three trends indicated that there was a viable concept at 1.8 GHz if new entrants targeted the handportable market. This concept was supported by U.S. research. For several years, Bellcore, the U.S. research arm of the local telephone operating companies, had been looking into handportable-based cellular radio networks. The researchers argued that a cellular network with vehicle sets and handportables on the same network would not serve either application in the most efficient way. They were also doing propagation studies in the 1.5- to 2.5-GHz range. This seemed to be the most likely range of frequencies where spectrum could be found in the United States for new handportable-based cellular networks. These ideas were brought to Europe when researchers from Bellcore were the guests at a GSM meeting in Berlin in 1985.

2.2 THE PERCEIVED ROLE OF PERSONAL COMMUNICATIONS

There were a number of other ideas for personal communications networks. In particular, the telepoint approach based on cordless telephones was generating a lot of interest. It was postulated that a series of public base stations at strategically located points would enable people to use their cordless telephone at home, at work, and via the public base stations as they traveled between the two. It was postulated that in the work location the private base stations might form a contiguous coverage over the work premises. This would require a system with a very fast hand-over. The small cells offered efficient use of radio frequency channels through a high reuse factor, which, in turn, would allow slightly wider bandwidths than contemplated for GSM. The short ranges obviated the need for sophisticated means of controlling multipath problems. These considerations led to a different strand of technology emerging for telepoint with little in common with GSM. These two strands of technology—PCS and telepoint—could be seen in the context of a wider picture of the public mobile radio market. In essence, mobile radio offers customers *mobility* for their telecommunications needs. The degree of mobility is one way to characterize the different mobile radio market offerings. At one extreme is the simple domestic cordless telephone, offering mobility within a 100-m radius. The other extreme is satellite telecommunications. One geostationary satellite offers mobility over one-third of the globe. The first costs very little by way of capital investment and negligible running cost for the cordless connection. The second involves a sizable investment for the terminal and a high running cost in terms of satellite charges. It is reasonable to suppose that the market comprises a full spectrum of user needs in respect to how important mobility is and how much it is worth. The ''Phones on the Move'' consultation document set out this spectrum of mobility offerings, as shown in Figure 2.1. The upward arrow shows the trend of building up increasing mobility from the simple cordless telephone in

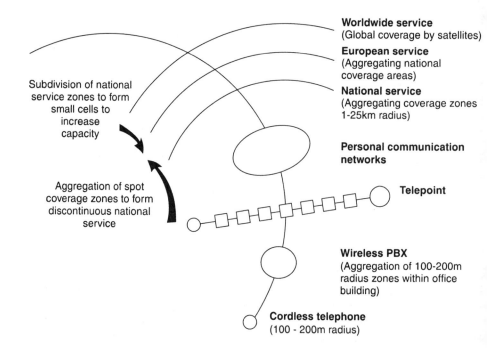

Figure 2.1 Geographical mobility differentiates markets for different mobile radio systems. (Reproduced from the U.K. government consultation paper, ''Phones on the Move,'' with the permission of the Controller of Her Majesty's Stationery Office.)

erms of telepoint. The downward arrow shows two things. The first is how a European ellular system is really a mosaic of interconnected national cellular systems. The second s the trend at the national level to cell–split, leading in time to a diminishing average ell size radius. It was postulated that PCNs could find their own unique space between elepoint and national cellular radio systems. Their capital investment cost may therefore also lie between the two.

This concept fit well with another aspiration of DTI, which was to find the means to bring effective competition in the local telephone loop. British Telecom had two enormous advantages over potential competitors. First, it had invested in its copper wire infrastructure over many decades, and the cost of much of that infrastructure had already been amortized. A competitor had to buy in at present-day costs and interest rates. But more significant was British Telecom's existing investment in ducts, which proved to be a significant barrier to new entrants. Radio access had the potential to leap over the duct investment barrier. The base station costs looked favorable against copper costs.

Radio technology had another advantage over copper wire for new entrants. Take a medium-sized town as an example. One radio base station with only one transmitter/ receiver channel pair would offer immediate service wherever a customer in that town happened to demand a service. Now say a second customer appeared at the other side of town. A second radio channel pair could be added, and so on. Eventually when there were enough customers, a second radio base station could be added to create, in effect, two radio cells. As the customer base built up, the investment in radio base stations would follow. In contrast, an extensive copper wire loop would have to have been laid before the first customer materialized. Thus, in investment terms, the radio access technology offers better investment and cash flow prospects.

The main disadvantage of radio access technology has been the cost of the subscriber equipment. The normal domestic telephone connected to a copper wire is very cheap. Cellular mobile radios have been relatively expensive. In the early days of radio technology, the general rule for making something cheap was to keep it simple. Modern semiconductor technology has rendered that rule almost obsolete. The new rule is to get high volumes of production. The higher the volume, the greater the complexity that can be contemplated. For example, one design of radio might cost a development laboratory $1 million. A second considerably more complex design may cost $10 million to develop. In either case, the development cost must be recovered in the sales. Suppose in both cases one million items a year are sold and the development costs are to be recovered in the first year of sales. The difference in price between the two items is only $9 in the first year and no difference thereafter. Yet there is an order-of-magnitude difference in development cost. This simple example brings out the importance of creating huge volumes to bring the technology within reach of ordinary customers. Thus, massive economies of scale are essential to realize the full potential of radio access technology for the local loop, though it is not necessary to bring the subscriber equipment to exactly the same level as the conventional domestic telephone since customers have the added value of some level of mobility.

2.3 CHOICE OF TECHNOLOGY

The identified need for high volumes led to the policy decision for DTI as to which technology should be used for PCN. Was it to be GSM or one of the telepoint technologies, such as DECT? Or should it be something quite different? The PCN competition was run in two rounds. The first was to seek industry reaction to the concept. The second was a runoff to decide who should receive licenses. In the "Phones on the Move" consultative document, DTI advocated GSM but did not preclude other possibilities. There was support from respondents for both the GSM technology and the cordless technology. In the second round DTI narrowed the choice down to GSM or DECT, for two reasons. The first was that a single standard was sought as a framework for competitive PCN operators. To narrow the choice prior to the competition for licenses made the task of a final selection of a single standard that much easier. Second, it made sense to build on the sentiments expressed by industry in the first round of consultation. There was support for both DECT and CT2 cordless technologies. But CT2 was a 900-MHz standard, whereas DECT was already envisaged for 1.8 GHz. In that way any European economies of scale for DECT would benefit the PCN application in the United Kingdom. Following the second round, there was complete industry support for the GSM standard. The choice was made.

The use of GSM technology was quite rational. It was the most advanced cellular radio standard in preparation and already had the support of a large number of countries. It had been designed with handportable radios in mind. It also kept open the option of serving the car market. But more important for any service aiming at setting itself on the path to a consumer market, it had assured European economies of scale behind it. While the 1.8-GHz frequency choice would require specific development, the complex digital processing VLSI chips would ride on the back of the GSM investment. It was also clear that industry could not have found the resources to develop a major new technology given their commitment to GSM.

There was a second way to rationalize the choice of GSM. The first public radio telephones had used VHF radio channels and frequency modulation. The next generation were the analog cellular systems that kept frequency modulation but moved to 900 MHz. The next generation maintained 900 MHz but changed the modulation to a digital system. What the U.K. PCN decision had done was to keep the same digital modulation system but move the frequency channels up to 1.8 GHz. Thus, PCN could be presented as a logical evolutionary progression. This concept converges with the view now widely held internationally that the next generation (so-called third-generation public systems) would maintain the 1.8 GHz but move to a more sophisticated modulation arrangement, perhaps by the turn of the century. This analysis was the basis of the United Kingdom presenting PCN to the rest of Europe as a generation 2.5, since the GSM was widely regarded as the second-generation cellular system.

2.4 THE WIDER IMPACT

The U.K. PCN initiative had a significant impact in the rest of the world. It coincided with the excitement over GSM, interest in Telepoint based on CT2, and the work on

DECT cordless technology. This put Europe in the forefront of mobile radio technology developments, which in turn led to pressure in the United States to move more rapidly with their own personal communications services initiative. However, within Europe the U.K. PCN initiative created some uncertainty for a while. What was a PCN, and how did it differ from 900-MHz GSM cellular radio? Some looked in vain for a new technological breakthrough. The rest of Europe did eventually settle for the concept with which they felt most comfortable. A PCN was simply a GSM network operating at 1.8 GHz. This was true as far as it went. It avoided awkward questions on competition to the local fixed infrastructure that remained a monopoly in most European countries. But such a conclusion missed the real significance of the PCN decision—the establishment of an uninterrupted road between today's cellular radio services for the business market and the well-off domestic subscriber to a point some time in the future when personal communications becomes a consumer mass market. At such a time it is inevitable that PCNs will compete with the local telephone network. The DTI decision put in place three essential conditions for this evolution. The first was to release enough radio spectrum (taking 1.8 GHz and 900 MHz together) so that a shortage of radio channels no longer constrained the growth of mobile radio subscribers. The second was to guide the development behind a technology where huge economies of scale seemed assured. The third was to maximize competitive drive by licensing 1.8-GHz operators in addition to the two existing 900-MHz operators.

What remains uncertain is the question of timing. Somewhere along the road to mass personal communications that now stretches out for the 1990s, industry expects the handportable to become the dominant subscriber unit (if it is not already). Further along the road the prices of handportable radio telephones are predicted to fall to a level within reach of most people. Not long after this a few PCN subscribers may decide to dispense with their fixed telephones. From that point, personal communications networks in various forms (900 MHz, 1.8 GHz, and various new technologies) will increasingly provide an access technology that is competitive with the local copper wire telephone loop.

Not that the fixed local telephone network will be a static target. The fixed network will evolve and likely exploit progressively its particular strength of wider bandwidth services, perhaps using fiber-optic connections to the customer as this becomes economical. Thus, the PCN will enter an environment of competitive technologies, securing its place through the provision of mobility at an affordable price to the mass market. This was the conceptual vision set out in the 1989 DTI "Phones on the Move" consultative document.

2.5 EXPECTATIONS OF SECOND-GENERATION SYSTEMS

While technical feasibility was not a sufficient condition for the emergence of PCNs, it was certainly a necessary one. Two aspects to the technical issue were closely related:

1. Could second generation air-interface design deliver the user capacity and range of services that would be demanded in the midterm future?

2. If this implied migration to higher-frequency bands, could terminal equipment costs be kept at the necessary modest level?

The early and sustained success of the analog cellular radio systems introduced into the United Kingdom in 1985 had established a view among those involved in charting the likely future trends in cellular service provision that there was a real need for substantial increase in user capacity in future systems. The mood was typified by pronouncements made by the cellular operators in the middle of 1984 at the first of several U.K. workshops held on cellular system technology. These workshops concentrated on the ways in which collaborative industry and university research might contribute to the evolution of the next generation of systems and to consolidate targets for the performance of such systems. A specific position was taken by the analog cellular operators that there would have to be at least an order of magnitude increase in user capacity in the next generation system to keep pace with the growing demand. At that time, work on what is now the GSM system (named after the Groupe Spécial Mobile of CEPT and subsequently ETSI [1–4]) was pursuing a wide variety of potential solutions to the implementation of digital techniques for cellular systems. These solutions were drawing on technology that had previously been actively deployed only in military applications, where costs constraints were of a different character. There seemed to be every reason to hope that the capacity of the new system would satisfy expectations. By the spring of 1987, however, views were consolidating around the combination of system parameters now defined as GSM. As a result there was an interesting combination of euphoria and concern: excitement that an advanced and sophisticated cellular radio system was emerging, giving a general sense of optimism that difficult technical solutions could be brought into the marketplace; concern that the target of "an order of magnitude increase" in capacity might not materialize in a single step to the second-generation system.

Perhaps the great achievement of the GSM system had been to establish a common system for Europe, not only in technical terms but also in having set in place the mechanism for future pan-European system developments.

2.6 THE PROMISE OF THIRD-GENERATION SYSTEMS

In parallel with these developments, the European Commission was defining the pan-European research program R&D in Advanced Communications Technologies in Europe (RACE), and a belief was growing that the desired capability of future systems would result from RACE efforts rather than from the implementation of GSM. Ideas began to formulate around a new universal mobile telecommunications system (UMTS), which would incorporate the findings of RACE research.

Equally, it was apparent by the latter half of 1988 that the RACE program was going to be slow in delivering its conclusions, and the historical experience of developing the GSM system suggested that a time scale of the end of the century for RACE implementation was not going to be significantly improved on. There was a further factor: the RACE community was predominantly orientated toward the broadband line communications aspirations of the fixed network operators. The feeling in RACE was that the key to the future lay with IBCN and the requirements of the personal communications community would be satisfied by essentially developing new capability in the fixed network with the radio access problem tackled separately. While this has a measure of truth in it, as evidenced by the emergence of universal personal telecommunications services (UPTS), in the fixed network, the cellular radio community was not convinced that the direction of RACE research was going to benefit its needs in the medium term. Members of that community did not see the mobile periphery as simply an add-on to the fixed network that would transparently deliver fixed network services. Rather they saw radio connection as having fundamental influence on fixed network developments, not the other way round.

This disquiet was reflected at the time in the preparatory work for RACE II, in which significantly greater emphasis was put on the radio-connected terminal problem. This was reassuring from the long term technical standpoint, but it did not address the time-scale problem. In fact, the reverse possibility was starting to emerge in that discussions were now taking place about the even longer-term prospects for global standards in what subsequently became known as future public land mobile telecommunications systems, or FPLMTS. CCIR established an interim working party, IWP 8/13, to define the requirements of FPLMTS and to consider the prospects of achieving them. This aspect was the result of growing concern within CCIR that regional standards no longer answered the long-terms needs of international telecommunications. Built into the FPLMTS concept was a further factor—the incorporation from the outset of integrated satellite-mobile system elements.

The result of these developments was to introduce further uncertainty: Was FPLMTS the same as UMTS or different? Was spectrum going to be earmarked for these new services? If so, when?

The telecommunications community clearly faced a dilemma: could the existing air-interface of GSM within the spectrum allocated in the 900-MHz band satisfy the demand for capacity and services that were anticipated until UMTS products become available? Or was it necessary to consider an interim measure? Clearly for some operators and administrations, the pressure on capacity was going to be more severe and nowhere greater than in the southeast of the United Kingdom. Not surprisingly, therefore, a ground swell of opinion began to emerge and gather strength that, although a completely new air-interface might be undesirable in advance of UMTS, more spectrum was going to be a priority and all the pointers were toward higher rather than lower sectors of the spectrum.

Besides the parallel political issues that entered into the argument, there were clearly technical considerations of fundamental importance to the operating and manufacturing industries at this time. the remainder of this chapter considers the main technical aspects that were perceived to be driving and constraining those developments.

2.7 THE TECHNICAL ENVIRONMENT

2.7.1 System Capacity

This issue of system capacity was central to the debate on the need for further evolution of the GSM system before the emergence of third-generation technology. A variety of different approaches were taken to estimate, on the one hand, what the requirements would be in the mid-1990s and, on the other hand, what GSM would be able to deliver. The arguments about demand focused on such parameters as density of population and levels of commercial activity, which were seen as the generators of peak demand. The business environment was regarded as one in which deterioration in service quality due to capacity saturation would not be tolerated. Basic population statistics for major conurbations in the United Kingdom, France, and Germany revealed densities per square kilometer of around 4,300 in London, 3,700 in Paris, and 2,100 in Hamburg, respectively. Given the concentration of commercial activity in the London and Paris areas, these were taken as the benchmarks for demand. By the end of the 1980s, the analog cellular system in the United Kingdom had been developed by both operators in response to user requirements. By sectoring cells and reducing cell sizes to maximize frequency reuse, the TACS system was estimated to be supporting around 2.5 to 3.0 erlang/km^2/MHz, which was felt to be, though not the limit of what could be achieved, a practical cost-effective ceiling. Also, this kind of capacity was possible only in limited coverage areas; it was not feasible to extend the techniques employed to general coverage.

Allowing for the anticipated migration of GSM into the spectrum occupied originally by TACS, it might be realistic to consider the capacity of 20 MHz allocated to GSM for comparison purposes. At the same level of utilization as the maximum achieved by TACS, this corresponds to 60 erlang/km^2, but if 75% of the urban center populace were to make a single 50-millierlang telephone call in the busy hour, the resulting traffic, 161 erlang, would approach three times the achievable capacity of the resource.

Further imponderables related to the extent to which general business telephone usage, while radio connected, would be able to take advantage of cordless technology rather than cellular. The unknown quantity then was the capacity of the emerging digital European cordless telecommunications (DECT) standard. Claims were beginning to surface in relation to the general-usage scenario for DECT and extreme situations, such as the traffic generated by financial dealing activity, which in a multistoried building at 0.2 erlang per user might produce tens of thousands of erlang/km^2. But if the business community became accustomed to cordless communications in the office environment, would this not increase their expectations of the cellular environment elsewhere? Moreover, concerns were increasingly being expressed about the kind of traffic that would be generated in the future; the plain old telephone service (POTS) was beginning to look like the tip of the iceberg when any comparisons were being made with the requirements of the fixed LAN community, where 10-Mbit/sec connections were commonplace.

What could be expected of GSM in this kind of demand environment was clearly the crucial question in relation to projected mid-1990s service provision.

The difficulties that this question posed to the technical specialists are exemplified by the relationship between capacity and frequency reuse. The problem is easily understood from elementary considerations of a generic cellular radio scheme with center illuminated cells. Of primary interest is the relationship between two ratios [5]: the minimum ratio of wanted received carrier to cochannel interference (C/I) and the ratio of the reuse distance (D) between cells in different clusters using the same frequency to the cell radius (R_c). The D/R_c ratio increases as the number of cells in a cluster increases.

The remaining factor relates to the fundamental mechanisms of radio propagation. In exact terms, these mechanisms are complex and the subject of continuing scientific study [6], but an approximate propagation law can be used to characterize the operating environment. By concentrating on the median field strength relationship with distance from the base station, a general relationship between field strength and distance is defined. For typical urban cells, this is usually assumed to be $1/d^4$, where d is the distance in question. Therefore, the required C/I determines the number of cells per cluster that are needed to obtain a defined D/R_c for a given propagation law.

For an analog system, the required C/I is a function of the deviation ratio of the FM transmission. In TACS this C/I requirement is about 10 dB, but allowing a margin for the difference between the median field strength and the field strength available for, say, 90% of the time due to the fading characteristics of the Rayleigh fading channel, a figure of 18 dB is more realistic. For a large cell environment in which all the cells are the same size and the $1/d^4$ law can be safely assumed, a cluster size of 7 emerges from the above relationships.

The relationship between cluster size and the ratio D/R_c is given by Parsons [5] as:

$$\frac{D}{R_c} = \sqrt{3N_c} \qquad (2.1)$$

where N_c is the number of cells in the cluster.

Now assuming the same propagation law holds for both wanted signals from a base station and the cochannel interference from surrounding clusters, it is seen that

$$\frac{C}{I} = \frac{R_c^{-4}}{\sum_{k=1}^{m} D^{-4}} \qquad (2.2)$$

where m clusters surround the cluster containing the cell of interest. Since six-sided polygons are used to represent cells, then six clusters will surround the cluster under consideration, that is, $m = 6$ [5]. Therefore,

$$\frac{C}{I} = \frac{1}{6\left[\dfrac{D}{R_c}\right]^{-4}} \qquad (2.3)$$

and

$$\frac{D}{R_c} = \sqrt{3N_c} = \left[6\frac{C}{I}\right]^{1/4} \qquad (2.3a)$$

that is,

$$N_c = \sqrt{\frac{2}{3}} \left[\frac{C}{I} \right]^{1/2} \qquad (2.4)$$

To increase capacity, terminals must be capable of operating in a significantly worse C/I. If a digital system could operate in a C/I of 3 dB, then, making the same allowances as before, the cluster size can be reduced to 4. If the four cells cover the same area as the original 7 cells then there will be 7/4 times as many channels available in the digital case, and capacity would have been increased proportionately.

This argument holds for reduction in the numbers of cells per cluster, assuming the same propagation law holds. To increase the system capacity beyond this requires reduction in cell size so that the available frequencies are reused more intensively, but this process cannot continue indefinitely without the propagation law changing to reflect the general reduction in the amount of scattering of the radio signal over shorter path lengths. There is a tendency in smaller cells for the propagation law to follow a $1/d^3$ relationship and, in extreme situations, when cells are so small that the terminals are within line of sight of the base stations, to $1/d^2$. Even in the $1/d^3$ environment, it becomes necessary to increase the number of cells per cluster once more to retain the same C/I due to the relationship in (2.4).

The above situation is considerably oversimplified, since in practice these difficulties with small cells are circumvented to some extent by sectoring cells and illuminating the sectors from their apexes, which is the technique employed in TACS to achieve these capacity figures. However, whatever techniques can be applied to TACS can also be used to effect with GSM, so the basic question remains, can GSM operate in a worse C/I environment than TACS and, if so, by how much? There is little to be gained from simply considering the numbers of users per unit bandwidth. TACS uses 25-kHz channel spacing and operates on a single channel-per-carrier basis so that eight users need 200 kHz of bandwidth. In GSM, the users access the radio resource by TDMA with eight users sharing each radio bearer; but these are spaced 200 kHz apart, corresponding to the eight users in TACS. There was general agreement, in the late 1980s, that GSM would perform well in much poorer C/I environments than analog systems, enhanced by such techniques as frequency hopping, which randomizes the interference to which terminals are subject, but estimates of capacity improvements in erlang/km^2/MHz ranged from 20% to 70% relative to TACS.

A further factor is the expectation that advances in speech-coding techniques will ultimately permit the introduction of a *half-rate* coder, which will enable each radio bearer to support sixteen rather than eight users. This will increase the final capacity of the GSM air-interface by a factor of 2. However, this relies on unpredictable progress in speech coder techniques and, subsequently, the displacement of full-rate equipment by half-rate terminals.

Notwithstanding the simplification employed in those arguments, it is easy to appreciate how some basic conclusions were drawn. In essence these were:

1. Increase in capacity beyond that achievable by GSM in the 900-MHz part of the spectrum would require a major advance in the fundamentals of digital transmission, bearing in mind that such strategies as the use of high-level modulation to achieve greater bits/sec/Hz efficiency, cannot be implemented to the same extent in a mobile/ personal communications environment as is possible in fixed-link applications. Therefore, further spectrum would be needed to meet capacity requirements.
2. This would probably mean progression to higher frequencies.
3. In terms of capacity alone, progress to a higher frequency could be advantageous since the $1/d^4$ law would hold for smaller cells, because of the relationship between the radio propagation characteristics and the transmission wavelength. However, this is only partially true since buildings and other environmental features stay the same size.
4. Small cells have other advantages as well in that the multipath propagation paths from terminals to bases and vice versa differ by smaller amounts, which means that the time dispersion effects will be less pronounced, higher bit-rates can be used, and channel equalization problems are reduced.

On the other hand, there is an obvious penalty in higher-frequency operation in that away from areas of high demand, coverage would require more base stations to compensate for the differing propagation characteristics—advantageous in urban centers, but a drawback in rural areas. Nevertheless, by 1989, the concept of microcells coexisting with larger cells had become widely accepted, and the consequences in terms of infrastructure costs attendant on utilization of higher frequencies were believed to be containable.

2.7.2 Hardware

In the initial stages of drawing up the GSM standard, many very different system architectures were considered, and some did not find wholehearted support because they appeared technically too complex for implementation in the intended time frame. When the "Phones on the Move" discussion document appeared in 1989, there was naturally some enthusiasm among the technical specialists for revisiting the more ambitious offerings for GSM. The counterargument to this was extremely persuasive, namely, that two air-interfaces, GSM and DECT would be receiving the full attentions of the semiconductor industry and any new system for personal communications would have to take advantage of this if costs were to be contained.

A number of elements in transceiver realization impinge on the cost issue. Broadly these elements are:

1. The RF circuitry, particularly the power amplifier part of the transmit section and the synthesizer, which sets the frequency stability of the RF part and determines frequency-hopping performance;

2. The signal processing elements, which need to be realized in a technology that is economical of dc power;
3. Battery technology.

It is immediately apparent that if GSM or DECT standards were targeted, the first and second elements would be the foci of much research and development anyway, so the RF components really represented the limiting factor. The DECT frequency allocation of 1.88 to 1.9 GHz would stimulate appropriate RF component development at a significantly higher frequency than GSM, but as a cordless system DECT developments would not be tackling the power amplifier requirement in the manner required for PCN. However, it was also apparent that at higher frequencies large-cell operation on the scale envisaged for GSM would not be appropriate, since the radio propagation characteristics favored small cells. Therefore, demands on power amplifier design would be significantly eased if a peak power level, perhaps 6 dB below that set for GSM, would result in readily realizable components.

REFERENCES

[1] Balston, D. M., "Pan-European Cellular Radio: Or 1991 and All That," *Electronics & Communication Engineering Journal*, Jan./Feb. 1989, pp. 7–13.

[2] Hodges, M. L., "The GSM Radio Interface," *Br. Telecom Technol. Journal*, Vol. 8, No. 1, Jan. 1990, pp. 31–43.

[3] Mouly, M., and M-B. Pautet, *The GSM System for Mobile Communications*, published privately, 1992.

[4] Steele, R., *Mobile Radio Communications*, London: Pentech Press, 1992.

[5] Lee, W. Y. C., *Mobile Cellular Telecommunications Systems*, New York: McGraw-Hill International, 1989.

[6] Parsons, J. D., *The Mobile Radio Propagation Channel*, London: Pentech Press, 1991.

Chapter 3

The Mobile Radio Channel

Adel Turkmani

Of all the radio-related research activities that have taken place over the years, those involving characterization of the radio propagation channel are among the most important and fundamental. The propagation channel is the principal contributor to many of the problems and limitations that beset mobile-radio systems. One of the major characteristics of radio channels is multipath propagation caused by terrain features and buildings. A line-of-sight path between transmitter and receiver seldom exists, and the principal modes of propagation are diffraction and scattering. The existence of multiple propagation paths, with different time delays, attenuations, and phases, gives rise to a highly complex, time-varying multipath transmission channel. For systems engineers to be able to determine optimum methods of mitigating the impairments caused by multipath propagation, it is essential for the transmission channel to be satisfactorily characterized. It is also important that the characterization should take into account the intended application of the communication system, that is, narrowband or wideband transmissions.

Future personal communication systems are expected to provide radio services in a large variety of locations, effectively wherever potential users are located. That covers a wide range of propagation channels, which are often classified according to the immediate surroundings of the mobile terminals, for example, dense-urban, urban, suburban, and rural areas. The terminals can also be outside or inside buildings. Therefore, channel characterizations should also include the effects of the different environments.

Multipath propagation causes rapid fluctuation in the received signal envelope, superimposed on which are slower variations in the mean-signal level owing to shadowing. In addition radio signals are also affected by the spatial separation between transmitters and receivers. This particular propagation component is known as the path loss. Figure 3.1 shows the three components of the received signal. It is clear that even over small distances, radio signals can experience large variations in their levels, and changes of 30

30

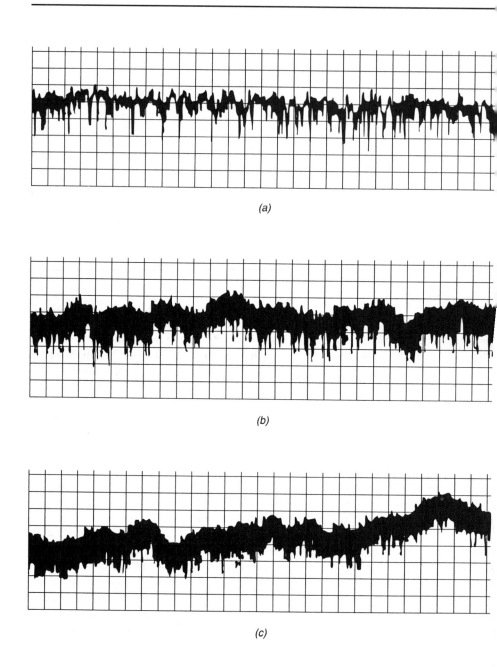

Figure 3.1 Processes that characterize the mobile radio channel: (a) short-term fading, (b) short-term fading and long-term fading, (c) short-term fading, long-term fading, and path loss.

o 40 dB are quite common. It is also evident that the waveform is a random process, hus implying a difficulty to deal with. This is a reality, and research workers have found ways to deal with this situation by extracting parameters that were found to be stationary. One of the most widely known of these parameters is the probability density function (PDF) of the envelope of the received signal.

The aim of this chapter is to present a detailed description of the mobile-radio channel and its effects. Propagation most appropriate for future personal communication systems will also be presented.

3.1 CHANNEL CHARACTERIZATION

Over the years, much attention has been devoted to the characterization of the multipath phenomena associated with mobile-radio signals. Particular emphasis has been placed on establishing an appropriate statistical distribution for describing the spatial variation of the complex random received signal. As early as 1952, Young [1] concluded, from a series of measurements in New York City at frequencies between 150 and 3,700 MHz, that over small areas, the signal amplitude variations could be closely represented by a Rayleigh distribution. This finding has been confirmed by countless experiments in several parts of the world in the intervening years. The Rayleigh distribution, however, only applied to the fast fading caused by multipath effects in the immediate vicinity of the mobile terminal.

As well as the search for the best statistical distribution, there have been several attempts to develop theoretical models for mobile-radio reception. The earliest of these was carried by Gilbert [2], who examined three different models, based on scattering rather than reflection. In all three models, the phases of the incoming waves were assumed to be independent and uniformly distributed in the interval $(0, 2\pi)$. One interesting finding by Gilbert was that for sufficiently large N, all three models are equivalent. In 1968, Clarke [3] published his well-known scattering model, based on Gilbert's second model. Clarke specified, a priori, a probability density function for the spatial angles of arrival, using an omnidirectional scattering model in which all angles are equally likely.

A common feature of all these models is that the incident waves are assumed to travel horizontally, that is, they are two-dimensional models. This basic assumption was questioned by Aulin [4], who argued that if this were true, no transmissions between a distant transmitter and a receiver on a street would be possible. Aulin proposed a generalization of Clarke's model so that the vertically polarized plane waves that make up the received signal did not necessarily travel horizontally; that is, a three-dimensional model. To obtain closed-form solution for the autocorrelation function and the power spectrum, however, Aulin had to assume a PDF for the spatial angle of arrival in the vertical plane, β. His chosen PDF was easy to handle mathematically and produced closed-form solutions, but it was not very realistic from the physical point of view. Recently Parsons and Turkmani [5] used Aulin's model but adopted a more realistic expression for the PDF of β based on data derived from experimental observations.

Transmissions from a base station reach the antenna of a mobile receiver via a number of different paths that exist owing to multiple scattering from buildings close to the mobile. It is often assumed [5], quite realistically, that the scatterers are uniformly located around the mobile, that is, that they form a circle around it. As the mobile moves, contributions from some scatterers disappear and are replaced by contributions from some new scatterers. Thus, a circle of scatterers always exists, and, despite the movement of the mobile, the angle of arrival can always be assumed to be uniformly distributed in the interval $(0,2\pi)$. By adoption of the generic multipath model proposed by Aulin, the scatterers may exist in three dimensions and therefore form a cylinder, rather than a circle, around the mobile. The dimensions of this cylinder depend on the environment surrounding the mobile. It has been shown by Turkmani and Parsons [6] that the effective radius scattering cylinder in urban and suburban areas is 60 to 70m. The effective radius will depend on the degree of urbanization, and its value is expected to vary considerably from one area to another. Parsons and Turkmani [5] also showed that the spatial angle of arrival in the vertical plane, β, has a maximum value in the range of 10 to 20 degrees. Thus, it can be concluded that the height of this assumed scattering cylinder is in the range 10 to 25m, which is obviously a realistic assumption.

In addition to the fact that the multipath fading signal has a Rayleigh distribution, a uniform phase distribution $(0,2\pi)$ and a uniform angle of arrival in the horizontal plane $(0,2\pi)$ with 10- to 20-degree spatial angle in the vertical plane, the spectrum, $A(f)$, of the received signal has a distinct bucket shape, which is given by

$$A(f) = \begin{cases} \dfrac{K}{4\pi f_m}\left[1 - \left(\dfrac{f}{f_m}\right)^2\right]^{-1/2} & |f| \leq f_m \\ \\ 0 & |f| > f_m \end{cases} \tag{3.1}$$

where K is a constant and f_m is the maximum Doppler shift. The multipath signal possesses another two stationary parameters, namely, the level crossing rate and the average fade duration.

3.1.1 Long-Term Fading

Over the years, the long-term fading has received less attention than the fast-term (Rayleigh) fading. It is well known that long-term fading is a random process caused by shadowing. Nevertheless, research workers have succeeded in showing that its PDF is stationary and can reasonably be approximated to the well-known log-normal distribution. In addition, the standard deviation of this term was foundi to be heavily dependent on the environment, and a value of 6 to 12 dB is often quoted in the literature. The fading rate for the log-normal component is not quite known, but it is linearly related to the frequency of operation and the speed of the mobile. Thus, the fading rate is linearly related to the maximum Doppler frequency, f_m. It is evident from Figure 3.1 that the fading rate of the log-normal component is much smaller than the fading rate of the multipath (Rayleigh) component. This may explain the reason why it is often taken to be a function, k, of f_m, and a value of k between 50 and 500 is often used [7].

In contrast to fast fading, slow fading is often modeled as a real (rather than complex) process. The shape of the spectrum of the log-normal component is not well known, and the literature contains no indication whether it should be of a similar shape to that given by (3.1), albeit at different fading frequencies. In reference [7] the log-normal component has been generated using a digital signal processing technique.

3.1.2 Wideband Propagation Characterization

Signals that have suffered multipath propagation constitute a set of randomly attenuated and phased replicas of the transmitted RF signal. The resultant sensed at the receiver is therefore the superposition of contributions from various individual paths. On that basis, it is reasonable to describe the mobile-radio channel in terms of a two-port filter with randomly time-varying transmission characteristics. The first general analytical treatment of time-variant linear filters was presented by Zadeh [8]. Bello [9] further developed the work of Zadeh and presented a general characterization of the channel, which he also applied to restricted classes of channels. The work of Bello was the basis of Demery's [10] work, in which a characterization of practical channels was presented. Kennedy [11] adopted an alternative approach to the time-variant filter description, using the point-scatterer mode. The observed characteristics lead to the conclusion that mobile-radio channels are strictly nonstationary, but in practice characterization proves extremely difficult unless stationarity is assumed over short distances of travel or short intervals of time.

A suitable statistical description of the channel has been proposed in the form of a two-stage model [9]. A small-scale characterization is first obtained over a period of time that is short in comparison to the period of the slow channel variations, so that the mean signal level appears sensibly constant. It is assumed that over this short period of time the prominent features are the same, that is, the significant scattering centers do not change. A large-scale characterization is subsequently obtained by averaging the small-scale statistics. This two-stage model was used by Cox [12], and the class of radio channel for which it applies is termed quasi-wide-sense-stationary (QWSS). A further simplification in the characterization of mobile-radio channels can be derived from the observation that contributions from scatterers with different time delays are uncorrelated. A description in terms of wide-sense-stationary-uncorrelated-scattering (WSSUS) statistics is therefore appropriate, and the channel can be envisaged as a continuum of uncorrelated scatterers in both time delays and Doppler shifts.

A time-domain description of the channel can be obtained by expressing the autocorrelation function R_w of the channel output in terms of the autocorrelation function of the input delay-spread function. The theory is well covered in the literature [9,10], and the relationship between the two autocorrelation functions, for WSSUS channels and zero time separation of observation, can be simplified to

$$R_w(t, t) = \int |Z(t - \tau)|^2 P(\tau) \, d(\tau) \qquad (3.2)$$

where τ is the time-delay variable and $z(t)$ is the complex envelope of the transmitted bandpass signal. Characterization in other domains is also possible using other autocorrelation functions. If $|z(t)|^2$ is an impulse function, then the autocorrelation function of the channel output is then described by the profile of the time distribution of received power, often termed the power time-delay profile. This description is valid provided $|z(t)|^2$ appears impulsive with respect to $P(t)$, this being satisfied if $z(t)$ exists for a much smaller time than the spread of multipath delays within the channel. Provided that the received signal has Gaussian statistics, then the channel behavior will be completely described by $P(t)$. The frequency-domain channel description is often specified in terms of the frequency correlation function $R_T(\delta f)$. It has been shown [9,10] that a separate measurement of $R_T(\delta f)$ is not required, since it is equal to the Fourier transform of $P(t)$. Many parameters have been used to represent the statistics of $P(t)$. The quantities *average delay* and *delay spread*, as defined by Cox [12], are the first and the second central moments, respectively, of the power delay profile. However, these two parameters alone are not sufficient to describe some of the important characteristics of the channel. In recognition of this fact, the use of two further parameters has been recommended, COST-207 [13], namely, *delay window* and *delay interval*. The two terms are used to describe the length of the impulse response and the distribution of energy within it.

3.1.3 Modeling of Wideband Propagation

For Gaussian wide-sense-stationary-uncorrelated-scattering (GWSSUS) channel, the power time-delay profile, $P(t)$, is sufficient to describe the behavior of the channel. Therefore, by relating the measured $P(t)$ to the local environments, it should be possible to develop a model for multipath channels assuming that only scattering takes place. For such a model to produce accurate results, a detailed environmental database is mandatory, but unfortunately such databases are extremely rare.

Because of the limited time-resolution capability of all practical measurement systems, the contribution to the power delay profile from echoes within a specific time-delay cell is the agglomeration of a large number of independent paths, each having a different phase. At UHF, small spatial changes in the location changes in the phase of each constituent path. Consequently, the vector addition of these randomly phased signals produces amplitude fluctuations in each time-delay cell within the power-delay profile for small spatial displacements. Intuitively, variations in the mean echo amplitudes due to large-scale changes in location are also expected, as a result of either range dependence or differing propagation environments. In view of these factors, it seems prudent to envisage the echo-path fluctuations as being separable into two distinct parts. The fluctuations can be then characterized in terms of small- and large-scale signal variations in an identical manner to that adopted for narrowband propagation studies. It has been shown by Turkmani [14] that the small-scale variations were found to be Rayleigh distributed. The large-scale variations, on the other hand, were best described by log-normal distribution with standard deviation of 3 to 6 dB.

In ascribing statistical distributions to the propagation time delays, several researchers have conjectured that these time delays form a Poisson sequence. However, after analyzing their measurements, they discovered that the distribution of the time delays was not consistent with a simple Poisson process. To overcome this finding, a modified Poisson sequence [15] was developed, which is similar to a Markov model. Although this refined model was better able to reproduce the features of the researchers' measurements, it is rather cumbersome to use. With the possible exception of computer models, the propagation time delays, for models based on tapped delay lines, will be fixed quantities. Although this conflicts with the physical channel consisting of a continuum of delays, it is a valid assumption for all channels of practical interest.

Also of importance are the correlation coefficients between amplitude fluctuations in neighboring time-delay cells. It has been shown in [14] that the small-scale fluctuations can be modeled as uncorrelated Rayleigh fading. In contrast to the small-scale fluctuations, the large-scale (log-normal) variations in neighboring (of $0.1\,\mu s$) time delays were found to have a correlation of 0.66. The correlation coefficients for separations greater than one time-delay cell are less than 0.5, and as such indicate that these large-scale fluctuations can be considered as weakly correlated, and for practical reasons they can even be assumed to be uncorrelated.

3.2 CELLULAR RADIO TOPOLOGY

Conventional mobile-radio telephony systems, based on large coverage areas for each base station, are not spectrally efficient and do not provide a high degree of flexibility for subscriber demands. These limitations were recognized as long ago as 1947, when Bell Telephone Company first began to consider a large-scale radio-telephone system based on a network of overlapping cells. As a result of many years of research and development, the first 800-MHz cellular system was introduced in the United States in 1979, and at about the same time a 450-MHz system was introduced in the Scandinavian countries. Subsequently, cellular systems have been successfully introduced in many other countries. The U.K. cellular network is based on a 900-MHz total access communication system (TACS), which is similar to the 800-MHz advanced mobile phone system (AMPS) adopted in North America. The other systems worldwide are the Nordic mobile telephone (NMT) and Japanese (NTT) systems.

The success of cellular systems is the result of the automatic capability of all present-day mobile-telephone networks. Each system provides full-duplex operation, automatic channel search, and dialing to and from the mobile station. As the level of sophistication of mobile-radio systems continues to increase, so must the sophistication of the techniques used to design and analyze those systems. In this section, a brief overview of cellular mobile-radio systems is presented; further detailed discussions of cellular systems can be found elsewhere [16]. However, in order to fully appreciate the operation and advantages of cellular-radio systems, it is necessary to understand the concepts of *frequency reuse, cell splitting*, and *call handoff*.

Figure 3.2 shows the basic structure of a cellular system. The service area to be covered is arranged into an appropriate network of contiguous radio cells. It is not possible to cover completely an area with circles, and so an idealized circular service area is represented by a hexagon. In practice, of course, service areas (cells) will not have a symmetrical shape because of the irregular nature of the terrain surrounding a base station. Each cell has a base station that utilizes an associated set of radio channels to effectively connect to any mobile unit located in the cell. The system uses two types of duplex channel. Control channels are used to transfer system-control data to and from the mobiles, and voice channels provide a link for speech or data transmissions and in-call supervisory tones. The base stations are connected, via the conventional land-line system or microwave links, to a mobile switching center (MSC). It is the MSC that controls the connection of the mobiles to each other and to the national and international telephone system.

3.2.1 Frequency Reuse

The geographical area shown in Figure 3.2(b) is covered by a pattern of cells, each having been assigned a set of control and voice channels. Identical sets of frequencies may be used by many base stations, providing there is sufficient spatial separation to ensure that cochannel interference is minimal. In practice, groups of adjacent cells share, between them, all the available radio channels, and the same frequencies can be reused in many parts of a service area. Therefore, cellular systems are often referred to as being *spectrally efficient*. In Figure 3.2(b) the channel groups are repeated in every seventh cell. Such an arrangement is known as a *seven-cell-repeat* pattern. Other repeat patterns are possible, and the size of the repeat pattern determines the separation distance between cochannel cells. Larger repeat patterns result in larger separation distances between cochannel cells; however, fewer channels can be assigned to each cell.

3.2.2 Cell Splitting

One of the major advantages of a cellular mobile-radio system is the inherent flexibility of the network to accommodate subscriber demands. By operating on a cellular basis, it is possible to increase the capacity of the system by simply splitting large cells into small cells, as illustrated in Figure 3.2(c). Such a system can employ cells of different sizes, depending on the density and distribution of traffic throughout the network. Usually, large

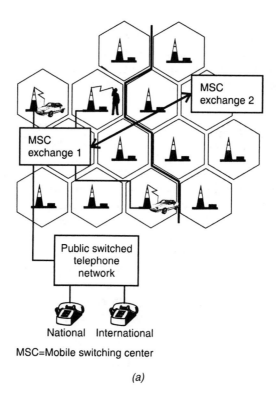

(a)

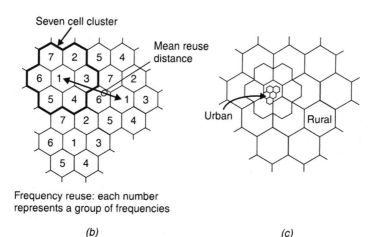

(b)   *(c)*

Figure 3.2 Cellular radio: (a) structure of the network; (b) cells based on an idealized hexagonal grid showing the principle of frequency reuse; (c) cell size selection.

cells would be used to provide service where there is a low subscriber density, and many smaller cells would serve areas of high subscriber density. In lightly populated areas, the cell diameter can be up to 30 km, whereas in the most densely populated areas, the cell diameter is only 2 km and smaller cells 1 km diameter are also in use.

3.2.3 Call Handoff

In a cellular network, it is important that the system provide the ability for mobiles moving across cell boundaries to maintain communication. This is achieved using a *handoff* technique. During a telephone conversation, the base station monitors the level of the received signal of the mobiles, and data are sent to each mobile to adjust its transmitter power as required. If the received signal strength falls below a predetermined lower limit, adjacent base stations are commanded to monitor the signal strength, and the call is "handed off" to the base station receiving the highest signal from the mobile. During the handoff process, signaling information is transmitted on the voice channel from the serving base site to the subscriber's equipment, giving new frequencies and supervisory information. The total handoff process is completed with minimal interruption in service.

3.3 PROPAGATION PREDICTION MODELS

A fundamental requirement in the design of land mobile-radio systems is knowledge of the mean received signal and its variability at all locations surrounding a base station site. With this knowledge, a systems designer can establish the coverage area and interference problems that may be associated with any base station. This information can be measured by radio surveys, but such surveys tend to be expensive, particularly since they would need to be repeated for each base station site. Consequently, it is advantageous for the systems designer to have available a computer-based system that can be used to predict both coverage areas and the likely effects of interfering stations. The basis of this system should be a reliable propagation model. In general, the propagation prediction models described in the literature are a mixture of empiricism and the application of analytical propagation theory [17].

Many of the available propagation prediction models have been subject to intense scrutiny over the years, resulting in a number of published review papers [18]. More recently, comparisons of some urban propagation models with carrier-wave (CW) measurements have also been presented in the European Research Committee's COST-231 program [19]. A comprehensive description of all these models and the supporting theory is also presented by Parsons [17]. Therefore, it was decided to limit the discussion in this section to the most widely used propagation prediction models, namely, the Okumura model and its extensions and those developed by the European research committee COST-231.

3.3.1 Okumura's Model and Its Extensions

The model has been developed from an extensive series of field trials, which were undertaken in and around Tokyo under the following conditions [20]:

- Frequencies from 100 MHz to 3000 MHz;
- Distances from 1 km to 100 km;
- Different terrain conditions:urban, suburban, rural with various degree of undulation;
- Effective base antenna height from 20m to 1000m;
- Vehicular antenna height from 1m to 10m;
- Other factors such as the orientation of streets and the presence of mixed land-sea paths.

The basis of the method is to determine the free-space path loss at a receiver located d_{km} from a transmitter and then add that value to the median attenuation, A_{mu}, in urban area over quasi-smooth terrain with a base station effective antenna height, h_{te}, of 200m and a mobile antenna height, h_{re}, of 3m. We can determine A_{mu} by using Figure 15 in [20]. The free space path loss at a frequency, say, f_{MHz}, can be determined using the following equation [17]

$$L_F \ (dB) = 10\log_{10}G_T + 10\log_{10}G_R - 20\log_{10}f_{MHz} - 20\log_{10}d_{km} - 32.44 \qquad (3.3)$$

where G_T and G_R are the gain of the transmitting and receiving antennas, respectively. Different correction factors can then be introduced to account for:

- Transmitting and receiving antennas not at reference heights;
- Transmission over non quasi-urban area, for example, suburban or rural;
- Orientation of streets;
- Presence of mixed land-sea paths.

Okumura produced different graphs that can be used to determine these correction factors. The Okumura model is probably the most widely quoted of the available models. It has come to be used as a standard by which to compare other models, since it is intended for use over a wide range of radio paths encompassing not only urban areas, but also different types of terrain [17].

In an attempt to make the Okumura model easy to apply, Hata [21] established empirical mathematical relationships to describe the graphical information given by Okumura. Hata's formulation is limited to certain ranges of input parameters and is applicable over quasi-smooth terrain. The mathematical expressions and their range of applicability are

$$\begin{aligned} L_{ha,Ur}(dB) = {} & 69.55 + 26.16\log_{10}f_{MHz} - 13.82\log_{10}h_{te,m} - a(h_{re,m}) \\ & + (44.9 - 6.55\log_{10}h_{te,m})\log_{10}d_{km} \end{aligned} \qquad (3.4)$$

where

$$150 \leq f_{MHz} \leq 1500$$
$$1 \leq d_{km} \leq 20 \qquad (3.5)$$
$$30 \leq h_{te,m} \leq 200$$
$$1 \leq h_{re,m} \leq 10$$

and $a(h_{re,m})$ is the correction factor for mobile-antenna height and is computed for a small- or medium-sized city and for a large city as

$$a(h_{re,m}) = (1.1\log_{10}f_{MHz} - 0.7)\,h_{re,m} - 1.56\log_{10}f_{MHz} + 0.8 \qquad (small\text{-}medium\ city) \qquad (3.6)$$

$$a(h_{re,m}) = \begin{cases} 8.29(\log_{10}1.54h_{re,m})^2 - 1.1 & f \leq 200\ MHz \\ 3.2(\log_{10}11.75h_{re,m})^2 - 4.97 & f \geq 400\ MHz \end{cases} \qquad (large\ city) \qquad (3.7)$$

The path loss for suburban areas is given by

$$L_{ha,Su}(dB) = L_{ha,Ur}(dB) - 2[\log_{10}(f_{MHz}/28)]^2 - 5.4 \qquad (3.8)$$

The path loss for open areas is given by

$$L_{ha,Op}(dB) = L_{ha,Ur}(dB) - 4.78(\log_{10}f_{MHz})^2 - 18.33\log_{10}f_{MHz} - 40.94 \qquad (3.9)$$

Hata's formulations, more commonly known as Hata's model, have enhanced the practical value of the Okumura model, since they are easily entered into a computer.

3.3.2 COST-231/Walfish/Ikegami Model

The European research committee COST-231 (evolution of land mobile radio) has created a combination empirical and deterministic model for estimating the urban transmission loss in the 900- and 1800-MHz bands known as the COST-231/Walfish/Ikegami model [22]. The model accounts for the free-space loss, the diffraction loss along the radio path, and the loss between the rooftops of the surrounding buildings and the mobile. It is mainly based on the models of Walfish and Bertoni [23] and Ikegami et al. [24]. Additionally, empirical corrections [25] were introduced in order to apply it to base stations and to match it to measurements considering street orientation and frequency [19]. The COST-231 model can be applied to radio paths in urban areas within the following ranges:

- Frequencies of 800 MHz to 2000 MHz;
- Distances of 200m to 5000m;

- Base station antenna heights of 4m to 50m;
- Mobile antenna heights of 1m to 3m.

The COST-231 model is composed of three terms:

$$L_{cost\text{-}231}(dB) = \begin{cases} L_F(dB) + L_{rts}(dB) + L_{mod}(dB) \\ L_F(dB) \end{cases} \qquad L_{rts}(dB) + L_{mod}(dB) \leq 0 \qquad (3.10)$$

where $L_F(dB)$ is the free-space loss, given by (3.3), $L_{rts}(dB)$ is the rooftop-to-street diffraction and scatter loss, and $L_{mod}(dB)$ is the multiscreen loss. $L_{rts}(dB)$ is based on Ikegami's model [24]:

$$L_{rts}(dB) = -16.9 - 10\log_{10}w + 10\log_{10}f_{MHz} + 20\log_{10}\Delta h_{mob,m} + L_{ori}(dB) \qquad (3.11)$$

where w is the street width and

$$\Delta h_{mob,m} = h_{roof,m} - h_{mob,m} \qquad (3.12)$$

is the difference between the building height, $h_{roof,m}$, and the mobile antenna height, $h_{mob,m}$. $L_{ori}(dB)$ is an empirical correction function that accounts for the street orientation and was evaluated from earlier measurements undertaken in Mannheim [25].

$$L_{ori}(dB) = \begin{cases} -10 + 0.354\ \phi & 0 \leq \phi < 35 \\ 2.5 + 0.075(\phi - 35) & 35 \leq \phi < 55 \\ 4.0 - 0.114(\phi - 55) & 55 \leq \phi \leq 90 \end{cases} \qquad (3.13)$$

where ϕ is the angle of incidence in degrees relative to the direction of the street. $L_{mod}(dB)$ is based on the Walfish and Bertoni model with some empirical corrections:

$$L_{mod}(dB) = L_{bsh}(dB) + k_a + k_d\log_{10}d_{km} + k_f\log_{10}f_{MHz} - 9\log_{10}b_m \qquad (3.14)$$

where b is the distance between the buildings along the path. The terms $L_{bsh}(dB)$ and k_a do not exist in the model of Walfish and Bertoni. They represent the increase of path loss due to reduced base station antenna height, h_{base}. The different terms introduced in (3.14) are given as follows:

$$L_{bsh}(dB) = \begin{cases} -18\log_{10}(1 + \Delta h_{base,m}) & h_{base,m} > h_{roof,m} \\ 0 & h_{base,m} \leq h_{roof,m} \end{cases} \qquad (3.15)$$

$$k_a = \begin{cases} 54 & h_{base,m} > h_{roof,m} \\ 54 - 0.8\ \Delta h_{base,m} & h_{base,m} \leq h_{roof,m},\ d_{km} \geq 0.5 \\ 54 - 1.6\Delta h_{base,m}d_{km} & h_{base,m} \leq h_{roof,m},\ d_{km} < 0.5 \end{cases} \qquad (3.16)$$

$$k_d = \begin{cases} 18 & h_{base,m} > h_{roof,m} \\ 18 - 15\dfrac{\Delta h_{base,m}}{h_{roof,m}} & h_{base,m} \le h_{roof,m} \end{cases} \tag{3.17}$$

where

$$\Delta h_{base,m} = h_{base,m} - h_{roof,m} \tag{3.18}$$

The term k_f is dependent on the transmission frequency and the degree of urbanization. For medium-sized cities and suburban centers with moderate tree density, k_f is given by

$$k_f = -4 + 0.7\left(\frac{f_{MHz}}{925} - 1\right) \tag{3.19}$$

For metropolitan centers, k_f is

$$k_f = -4 + 1.5\left(\frac{f_{MHz}}{925} - 1\right) \tag{3.20}$$

In the absence of detailed data about building structure, COST 231 recommends employing these default values:

- $b = 20$m to 50m
- $w = b/2$
- $h_{roof,m} = 3$ (number of floors) + $roof$
- $roof = 3$m for pitched; 0m for flat
- $\phi = 90$ degrees

The COST 231 model has been tested recently by Low [19]. A number of measurements were undertaken in the cities of Mannheim, where the land cover can be characterized as homogeneous, and in Darmstadt, a city with unhomogenous built-up structure, irregular streets, and small terrain undulations. Low reported that the comparison with measurements resulted in a good path-loss estimation for base antennas installed above the roof tops of the adjacent buildings. The mean error is within ±3 dB and the standard deviation within 5 to 7 dB.

3.4 PROPAGATION FOR PERSONAL COMMUNICATIONS

The full economic potential of personal communications systems is likely to be achieved only if there is extensive use of hand-held terminals, which should provide services to potential users, almost everywhere; outside and inside buildings, from dense-urban to fairly open areas. In addition, the users can be stationary or passengers in high-speed trains. A better understanding of the propagation into and within buildings and in microcells s therefore essential. In this section, a concise description of these propagation scenarios will be given.

1.4.1 Propagation Into Buildings

The move toward personal communications has led to the realization that not enough is known about radio propagation either into or within buildings. In this context, *into* is used to identify the propagation scenario where a base station, often located on a hilltop ite or rooftop of a high building, communicates with a radio receiver that is inside another building. Propagation models that adequately describe the signal in open and urban areas re no longer adequate, since there will be a building-penetration loss associated with the indoor environment. This additional loss will depend on a large number of factors having arious degrees of importance. Among them are the transmission frequency, the distance etween, the transmitter and the receiver, the building construction material and the nature f the surrounding buildings.

Two definitions of building penetration loss are found in the literature. Rice [26] efines the penetration loss as the difference between the average received signal strength easured over a small area within a building and that measured outside at street level ound the perimeter of the building. The definition used by Durante [27] compares the gnal strength inside and outside the building at the same height at various floor levels. he availability of propagation models designed to predict the signal level at street level akes Rice's definition more practical and useful because it allows the prediction models to be extrapolated from the outside to the inside of buildings. In addition to the penetration loss, systems designers are also interested in learning about the received signal variability and the effects of building height, conditions of transmissions, construction materials, and the frequency of operation. Several research activities that deal with these aspects have been reported in the open literature [26–30]. The general conclusions are:

- The small-scale signal variation is Rayleigh distributed [28].
- The large-scale signal variation is log-normally distributed with a standard deviation related the condition of transmission and the area of the floor.
- For no-line-of-sight transmissions, the standard deviation is approximately 4 dB [28].
- For partial- to complete-line-of-sight conditions, the standard deviation increases to 6 dB to 9 dB.
- The penetration loss decreases at higher frequencies [26,28,29]: 18 dB at 441 MHz, 14.5 dB at 900 MHz, 13.4 dB at 1800 MHz, and 12.8 at 2300 MHz.
- The rate of change of penetration loss with height is approximately −2 dB per floor [27,28].

However, it is necessary to remember that most of the outdoor propagation models have been developed for large cells, whereas for personal communications, the suitable cell diameter could be even less than 500m. Therefore, those models cannot be trusted when used for indoor environments without further investigations. In addition, predicting first the signals outside the building of interest and then, from that result, determining the signals inside the building yields an inevitable reduction in accuracy. Therefore, prediction of the path loss for radio transmissions into buildings may be more accurate if it has been undertaken directly and not merely as an extension of outdoor propagation models. A similar approach was adopted by Barry and Williamson [30] to analyze measurements undertaken in New Zealand at 851 MHz. Toledo [29] has obtained direct modeling of propagation into buildings at 900, 1800, and 2300 MHz.

3.4.2 Propagation Within Buildings

The term *propagation within buildings* is often used to identify the propagation scenario when both transmitter and receiver are inside the same building. Most of the research activities that have been reported in the literature [31–33] are mainly concerned with the investigation of the small- and large-scale statistics of the received signal and the variability of the floor's mean signal level. The significant conclusions related to propagation within buildings are as follows:

- The small-scale signal variations are Rayleigh distributed.
- The large-scale signal variations are, reasonably, log-normally distributed with a standard deviation value of about 16 dB [31].
- The rate of change of mean signal per floor (at 1800 MHz) was approximately 8.3 dB/floor; 6.1 dB/floor for floors below the transmitter, and −10.4 dB/floor for floors above the transmitter [31].

Figure 3.3 shows the normalized floor mean signal level against number of floors separating the receiver and transmitter.

Of importance also is the dependence of the overall signal coverage on the position of the transmitter inside the same building. It has been shown by Turkmani and Toledo [31] that, by locating the transmitter in a very large room in the middle of the building, the signal coverage can be increased substantially. This is desirable in a one-cell-per-building signal coverage. For a multicell-per-building radio system, other aspects beside signal coverage should also be considered. One of the earliest approaches to statistical modeling of propagation totally within buildings was reported by Alexander [32], who stated that the path loss within buildings at 900 MHz can be predicted using the simple distance/power law. Motley and Keenan [33] have also undertaken a series of indoor measurements at 900 and 1700 MHz in a building of standard steel-frame construction with brick external walls and plasterboard internal partitions. Motley and Keenan modeling results have shown that a better fit to the experimental data can be achieved by introducing

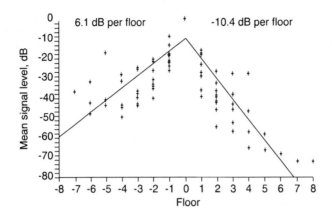

Figure 3.3 Normalized floor mean signal level against number of floors separating the receiver and the transmitter.

to the Alexander model a correction factor (F_{floor}) representing the signal attenuation per floor. Modeling of propagation totally within buildings has been reported more recently by Toledo [29]. The measurements were undertaken at three different frequencies (900, 1800, and 2300 MHz) and in four different buildings at the University of Liverpool.

3.4.3 Microcellular Propagation

Although GSM is designed to have a better spectrum efficiency, the need for personal communications in urban areas is expected to exceed the capacity that can be achieved by using conventional cellular planning methods. Even capacity improvement by cell splitting has practical limits due to antenna site constraints.

A promising idea that has emerged over the last few years and that has the potential to solve this serious problem is the microcellular approach [34–41]. It implies decreasing the cell size in an attempt to increase frequency reuse, leading to a much higher system capacity. Microcellular radio, however, has some drawbacks, principally the system complexity (large number of base stations), the increased number of handovers that occur during one phone call, and the severe effects of cochannel interference. It is, therefore, justifiable to use microcellular radio only where it is most needed. With this in mind, we can see that future cellular systems will contain large cells as well as microcells. The potential demand and its geographical location will dictate the cell size. For example, rural areas are likely to be served by large cells, suburban areas and some urban areas by small cells, and, more likely, dense urban areas by microcells.

Extensive series of microcell propagation measurements have been undertaken in different parts of the world in typical areas (i.e., areas where microcellular radio is expected to be operated). The experimental results are then analyzed to provide an appropriate empirical modeling of the microcellular propagation aspects. Figure 3.4(a) shows a typical received signal envelope in main streets, that is, where the transmitter is located. Figure 3.4(b) shows the typical received signal in side streets. Because a line-of-sight path transmission seldom exists in side streets, the received signal exhibits, as expected, more variability (the number of fades increases). The fast-fading component of the received signal can be obtained by normalizing the measured signal by its running mean using the moving average technique with a window width of 20 wavelengths. It has been shown in [34] that the fast-fading component of the received signal in both main and side streets can be accurately modeled by Rice distribution.

It is also clear from Figure 3.4(b) that the signal drops sharply (by around 20 dB) when the receiver is moved for a short distance along the side street, away from the main street. This drop, however, seems independent of the antenna height and the frequency of transmission [34]. The distance over which the signal level drops by 20 dB is related to the surroundings. If the transmission to the side street is blocked by multistoried buildings (the typical case in city centers), the signal drops by 20 dB over a distance of around 25m to 30m. The distance increases to 45m to 50m if the transmission is blocked by one-story buildings. The local mean variability in main streets has attracted more

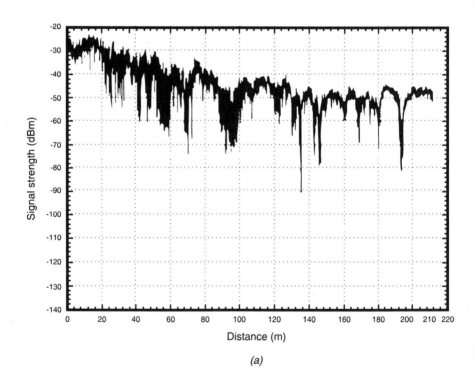

(a)

Figure 3.4 Measured signal strength: (a) main street, (b) side street.

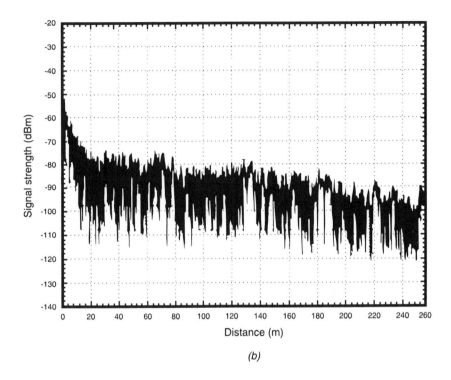

(b)

Figure 3.4 (continued).

attention by research workers than the small-scale variations, because they affect the dimension of the coverage areas in microcell. The local mean level is often modeled by a power/distance factor. It has been shown in [35] that the local mean in main streets has two power/distance factors, n_1 and n_2, with a turning point (Figure 3.5). A summary of the local mean slope values is presented in Table 3.1. It is evident that the values of n_1, n_2, and the turning point exhibit wide variability ranges.

3.4.4 Modeling of Microcells

In addition to the empirical modeling described, research workers have developed analytical models that are mainly based on the ray-tracing technique. The basic idea underlying this technique is that radio waves from the transmitter antenna are reflected by the ground and by buildings. The received signal at the mobile therefore consists of a direct ray, a ray reflected by the ground, and rays reflected by buildings. These components can be calculated using approximate mathematical formulas [36]. When N rays are incident on the mobile antenna, the received signal is the vector sum of the fields due to these rays.

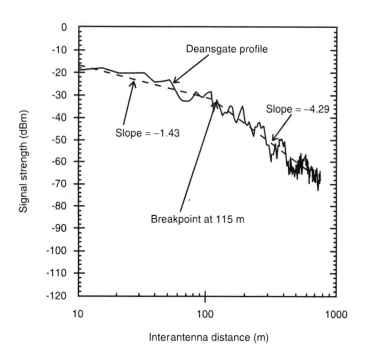

Figure 3.5 Typical signal strength profile measured in the main street showing dual path-loss characteristic and turning point.

The building walls are assumed to be smooth, contiguous, infinitely high, imperfect dielectrics with complex planar surfaces. The ground surface is assumed to be flat and perpendicular to the walls. In reference [36], the signal energy in main streets is assumed to travel to the mobile via the following paths:

- Line of sight;
- Ground reflection;
- Single- and multiple-wall reflections.

Reflections are considered to a depth of 4, which gives a total of 10 rays. For side-street computation, the signal energy arrives at the mobile via the following paths:

- Line-of-sight, which ceases to exist farther into the side street;
- Ground reflection, which also ceases to exit farther into the street;
- Wall-reflected rays;
- Wall-diffracted rays;
- Wall-diffracted-reflected rays;
- Wall-reflected-diffracted rays;
- Wall-reflected-diffracted-reflected rays.

Table 3.1

Summary of Pathloss Factors from Previous Measurement Programmes

Reference	Frequency (MHz)	Location	Mobile Antenna Height (m)	Base Antenna Heights (m)	Pathloss Characteristic		
					n_1	n_2	d_{brk}
Turkmani et al.	900 1500 1800	Liverpool, U.K.	1.4	7.0 14.0 21.0	2.20 2.17 2.01		
E. Green	905.15	London, U.K.: Harley St... Edgware Rd... Maida Vale... Marylebone Rd. Abbey Rd...(all urban)	1.5	5.0	2.14 1.70 1.72 1.85 1.74	4.46 8.30 9.80 7.41 9.19	271.0 m 207.0 m 316.0 m 273.0 m 280.0 m
K.L. Blackard	1900	San Francisco and Oakland, U.S.A. (urban/sub-urban)	1.7	3.7 8.5 13.3	2.18 2.17 2.07	3.29 3.36 4.16	159.0 m 366.0 m 573.0 m
P. Harley	870.15 (and 1800 but results not given)	Melbourne, Australia (urban)	1.5	5.0 9.0 15.0 19.0	1.15 0.74 0.20 -0.48	1.01 1.01 1.25 1.88	148.6 m 151.8 m 143.9 m 158.3 m
Xia et al.	1850 1937 (and 900 MHz band but not reported)	San Francisco Bay, U.S.A. Rural (1850) Sub-urban (1937) Urban (1937)	1.6	8.7 3.2 8.7	1.2 1.0 1.2	4.1 4.8 13.0	343.6 m 132.1 m 1000.0 m
Berg J-E et al.	870	Stockholm, Sweden (urban)	2.0	5.0	2.0 2.0 2.0	3.2 4.6 6.2	143.0 m 189.0 m 209.0 m

50

This gives a total of 12 rays. Parsons et al. [37] has shown that the model can be improved further if each wall-reflected ray is assigned a certain probability of occurrence in order to indicate the presence or absence of buildings along streets as well as the imperfection of the ground and building surfaces.

In addition to the transmitter power, Arowojolu and Turkmani [36] investigated the effects of the base station antenna height, directivity, and downtilt as a means of controlling the coverage area of a microcell. The investigation is software-based and employs a signal strength prediction algorithm. The computations have been undertaken at 1800 MHz with a transmitter power of 1 mW. The side street is assumed to be 200m away from the base station in the main street. To study the effects of the base station antenna height variation, the three height values of 5m, 10m, and 15m were used. A mobile antenna height of 1.5m was used for all calculations. To investigate the effects of antenna directivity, two different antennas were used. Table 3.2 summarizes the electrical characteristics of these antennas. Mechanical downtilts of 0, 30, 60, and 90 degrees were applied to the 15-element Yagi-Uda antenna to investigate the effect of beam tilting. A quarter-wave monopole at the center of a perfectly conducting ground plane served as the mobile antenna.

Figure 3.6 shows the signal strength profiles obtained when mechanical downtilts of 0, 30, 60, and 90 degrees are applied to the 15-element Yagi-Uda antenna. It is interesting to note that close to the base station (≤45m), the 30 degree profile is higher than the 0 degree profile. Farther into the street, however, the 30 degree profile is about 16 dB less than the 0 degree profile for the first 20m and is subsequently 20 dB lower farther into the street. For 90 degree tilt, the signal drops further, almost 45 dB, into the street, and the profile exhibits a sharper attenuation rate. This characteristic is due to the fact that when a *highly directive* antenna is tilted, its main lobe is tilted and it illuminates an area closer to the base station with increased signal strength but with lower signal strength in regions away from it. The faster decaying rate on the periphery of the cell helps to reduce interference in neighboring cells using the same channel set.

3.5 CONCLUDING REMARKS

It should be clear that the transmissions of wideband signals over mobile-radio channels are subjected to severe frequency-selective fading, due to the multipath nature of such

Table 3.2
Electrical Characteristics of Antennas

Antenna Type	Gain (dBi)	Vertical Beamwidth (deg.)	Horizontal Beamwidth (deg.)
Yagi-Uda array, 15 elements	13.74	30	30
Halfwave dipole	2.16	78	360 (omni)

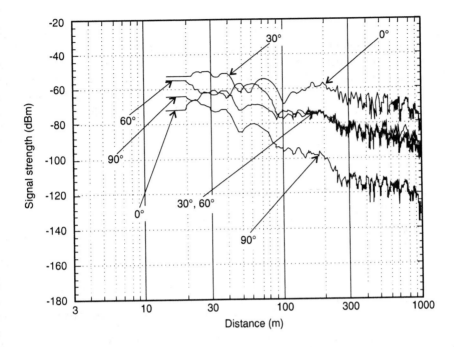

Figure 3.6 Signal strength profiles in the main street for different tilt angles of 0, 30, 60, and 90 degrees of the base antenna. Results have been obtained using the 15-element Yagi array antenna.

channels. It is well known that fading seriously affects the transmissions, and acceptable performance is therefore not possible unless steps are taken to counteract the channel imperfections. Fortunately, a large body of research work concerning wideband transmissions has been undertaken not only to study the characteristics of the received signal but also to develop techniques to minimize the effect of impairments associated with the channel. It is expected that future wideband systems will cope with these impairments reasonably well. These systems will use speech and channel coding, interleaving, and channel equalization. It is also expected that diversity techniques will be used in most future systems. This is due to the fact that they have been successfully utilized in narrowband signal transmissions. The use of diversity in wideband transmissions has not been given much attention until recently [42,43]. Nevertheless, they have proved that diversity is a promising technique for improving the performance of wideband transmission systems.

Since the main contributor to the performance degradation of wideband transmissions is frequency-selective fading, it is anticipated that the power-delay profile of the received signal will be an important parameter. Therefore, it should be, directly or indirectly, considered in any diversity system. In practice, however, to implement diversity using a direct measurement of power-delay profile may prove to be complex and costly. This complexity can be simplified with the appreciation that in a diversity system exact evaluation of the signal parameter is not often required. In fact, only *relative* evaluation of the desired parameter is required. For example, in a two-branch diversity system, it is sufficient to establish, for a given period of time, which of the two channels exhibits less severe frequency-selective fading for its output to be selected or heavily weighted.

REFERENCES

[1] Young, W. R., "Comparison of Mobile Radio Transmission at 150, 450, 900 and 3700 MC," *Bell Syst. Tech J.*, 31, 1952, pp. 1068–1085.
[2] Gilbert, E. N., "Energy Reception for Mobile Radio," *Bell Syst. Tech. J.*, 44, 1965, pp. 1779–1803.
[3] Clarke, R. H., " A Statistical Theory of Mobile-Radio Reception," *Bell Syst. Tech. J.*, 47, 1968, pp. 957–1000.
[4] Aulin, T., "A modified Model for the Fading Signal at a Mobile Radio Channel," *IEEE Trans.*, VT-28 (3), 1979, pp. 182–203.
[5] Parsons, J. D., and A. M. D. Turkmani, "Characterisation of Mobile Radio Signals: Model Description," *IEE Proc. I, Commun., Speech & Vision*, 138 (6), 1991, pp. 549–556.
[6] Turkmani, A. M. D., and J. D. Parsons, "Characterisation of Mobile Radio Signals: Base Station Crosscorrelation," *IEE Proc. I, Commun., Speech & Vision*, 138 (6), 1991, pp. 557–565.
[7] An, J. F., A. M. D. Turkmani, and J. D. Parsons, "Implementation of a DSP-Based Frequency Non-Selective Fading Simulator," *IEE 5th Int. Conf. on Radio Receiver and Associated Systems*, Conf. Publ. No. 325, Cambridge, 1990, pp. 20–24.
[8] Zadeh, L. A., "Frequency Analysis of Variable Networks," *IRE Proc.*, 38, 1950, pp. 291–299.
[9] Bello, P. A., "Characterization of Randomly Time-Variant Linear Channels," *IEEE Trans.*, CS-11, 1963, pp. 360–393.
[10] Demery, D. A., "Wideband Characterisation of UHF Radio Channels in Urban Areas," Ph.D. Thesis, Department of Electrical Engineering and Electronics, University of Liverpool, 1989.
[11] Kennedy, R. S., *Fading Dispersive Communications Channels*, Wiley-Interscience, 1969.
[12] Cox, D. C., "Delay Doppler Characteristics of Multipath Propagation at 910 Mhz in a Suburban Mobile Radio Environment," *IEEE Trans.*, Vol. AP-20, No. 5, 1972, pp. 625–635.
[13] "Propagation Data and Prediction Methods for the Terrestrial Land Mobile Service Using the Frequency Range 30 MHz to 30 GHz," *Report 567-3 (Mod F), CCIR XVIIth Plenary Assembly*, Dusseldrof, 1990.
[14] Turkmani, A. M. D., D. A. Demery, and J. D. Parsons, "Measurement and Modelling of Wideband Mobile Radio Channels at 900 MHz," *IEE Proc. I, Commun., Speech & Vision*, 138 (5), 1991, pp. 447–457.
[15] Hashemi, H., "Simulation of the Urban Radio Propagation Channel," *IEEE Trans.*, VT-28 (3), 1979, pp. 213–225.
[16] MacDonald, V. H., "The Cellular Concept," *Bell Syst. Tech. J.*, Vol. 58, No. 7, 1979, pp. 15–41.
[17] Parsons, J. D., *The Mobile Radio Propagation Channel*, Pentech Press Ltd., 1992.
[18] Delisle, G. Y., et al., "Propagation Loss Prediction: A Comparative Study With Application to the Mobile Radio Channel," *IEEE Trans.*, VT-34, 1985, pp. 86–96.
[19] Low, K., "Comparison of Urban Propagation Models With CW Measurements," *COST 231 TD (92) 44*, Leeds, 1992.
[20] Okumura, Y., et al., "Field Strength and Its Variability in VHF and UHF Land-Mobile Radio Service," *Review of the ECL*, 16, 1968, pp. 825–873.
[21] Hata, M., "Empirical Formula for Propagation Loss in Land Mobile Radio Services," *IEEE Trans.*, VT-29, No. 3, 1980, pp. 317–325.

[22] COST, "Urban Transmission Loss Models for Mobile Radio in the 900 and 1800 MHz Bands," *COST 231 TD (91) 73*, 1991.

[23] Walfish, J., and H. L. Bertoni, "A Theoretical Model of UHF Propagation in Urban Environments," *IEEE Trans.*, AP-38, 1988, pp. 1788–1796.

[24] Ikegami, F., et al., "Propagation Factors Controlling Mean Field Strength on Urban Streets," *IEEE Trans.*, AP-32, 1984, pp. 822–829.

[25] Rathgeber, R., F. M. Landsdorfer, and R. W. Lorenz, "Extension of the DBP Field Strength Prediction Programme to Cellular Mobile Radio," *IEE ICAP Conf. Proc.*, 333, 1991, pp. 164–168.

[26] Rice, L. P., "Radio Transmission Into Buildings at 35 and 135 MHz," *Bell Syst. Tech. J.*, 38, No. 1, 1959, pp. 197–210.

[27] Durante, J. M., "Building Penetration Loss at 900 MHz," *Proc. IEEE VT Conf.*, 1973, pp. 1–7.

[28] Turkmani, A. M. D., J. D. Parsons, and D. G. Lewis, "Measurement of Building Penetration Loss on Radio Signals at 441, 900 and 1400 MHz," *J. of IERE*, Vol. 58, No. 6 (supplement), 1988, pp. S169-S174.

[29] Toledo, A. F., "Narrowband Characterisation of Radio Transmissions Into and Within Buildings at 900, 1800 and 2300 MHz," Ph.D. Thesis, University of Liverpool, 1992.

[30] Barry, P. J., and A. G. Williamson, "Statistical Model for UHF Radio-Wave Signals Within Externally Illuminated Multistorey Buildings," *IEE Proc. Part I*, 138, (4), 1991, pp. 307–318.

[31] Turkmani, A. M. D., and A. F. Toledo, "Radio Transmission at 1800 MHz Into and Within Multistorey Buildings," *IEE Proc. Part I*, 138, No. 6, 1991, pp. 577–584.

[32] Alexander, S. E., "Radio Propagation Within Buildings at 900 MHz," *Electronics Letters*, 18 (21), 1982, pp. 913–914.

[33] Motley, A. J., and J. M. P. Keenan, "Personal Communication Radio Coverage in Buildings at 900 MHz and 1700 MHz," *Electronics Letters*, 24 (12), 1988, pp. 763–764.

[34] Turkmani, A. M. D., et al., "Microcellular Radio Measurements at 900, 1500 and 1800 MHz," *IEE 5th Int. Conf. on Land Mobile Radio*, IEE Conf. Publ. 315, Warwick, 1989, pp. 65–68.

[35] Green, E., "Path Loss and Signal Variability Analysis for Microcells," *IEE 4th Int. Conf. on Personal and Mobile Communications*, 1989, pp. 38–42.

[36] Arowojolu, A. A., and A. M. D. Turkmani, "Controlling the Coverage Area of a Microcell," *IEE ICAP Conf. Proc.*, 370, 1993, pp. 72–75.

[37] Parsons, J. D., A. M. D. Turkmani, and M. Khorami, "Microcellular Radio Modelling," *IEE 6th Int. Conf. on Mobile Radio and Personal Communications*, Warwick, 1991, pp. 182–190.

[38] Blackard, K. L., et al., "Path Loss and Delay Spread Models as Functions of Antenna Height for Microcellular System Design," *IEEE VT Conf.*, 1992, pp. 333–338.

[39] Harley, P., "Short Distance Attenuation Measurements at 900 MHz and 1.8 GHz Using Low Antenna Heights for Microcells," *IEEE J.*, SAC-7, No. 1, 1989, pp. 5–11.

[40] Xia, H. H., et al., "Radio Propagation Measurements and Modelling for Line-of-Sight Microcellular Systems," *IEEE VT Conf.*, 1992, pp. 349–354.

[41] Berg, J-E., R. Bownds, and F. Lostse, "Path Loss and Fading Models for Microcells at 900 MHz," *IEEE VT Conf.*, 1992, pp. 666–671.

[42] Acampora, A. S., and J. H. Winters, "A Wireless Network for Wide-Band Indoor Communications," *IEEE J. Select. Areas Com.*, Vol. SAC-2, No. 5, 1987, pp. 796–805.

[43] Chaaban, M., "Performance Evaluation of a TDMA Digital Mobile Radio System," Ph.D. Thesis, Department of Electrical Engineering and Electronics, University of Liverpool, 1993.

Part II
Implementation: Some Real and Emerging Systems

Chapter 4

The TDMA Approach: DECT Cordless Access as a Route to PCS

Hans van der Hoek

As analog cellular radio is replaced by second-generation digital systems such as (in Europe) GSM and DCS1800, the attention of the telecommunications community is turning toward the specification of future mobile communications systems aimed at increasingly personalized services. However, an alternative approach to personal communications is also emerging, termed personal communications services, or PCS. This approach is based on the combined strengths of two entities: the existing telephone network (either private or public) and a radio access part.

In this chapter, we explore this approach in relation to cordless access that conforms to the DECT standard.

4.1 PCS

The aim of PCS is to provide personalized voice, data, image, and video communications services that can be accessed regardless of location, network, and time. The PCS concept includes terminal mobility, personal mobility, and service mobility.

PCS can bring many benefits to its users. Personalizing the communications services is in effect an increase in the efficiency of the telecommunication services to deliver calls to its users. Current telephone networks supply services at places where people are expected to be. For example, in a private branch exchange (PBX) network, calls are delivered to desks, and it is far from certain that the person for whom a call is intended is at his or her desk.

Personalizing networks means, in effect, that users can access services wherever they are and, hence, that services are made available where the users are, at any moment, at any place.

Users may apply PCS for whatever benefits they want to derive from it, like improving convenience levels, business performance levels, competitiveness, or service levels. PCS may also be applied for reducing the costs of business operations.

If a user requires service access only in a limited area, for example, merely in and around the home, a system consisting of a single radio cell that can offer several access channels (i.e., a single-cell multiuser, or SC/MU, system) will probably meet that user's requirements. An SC/MU system is a simple "radio tail" of the network, delivering local mobility.

If services have to be accessible in larger areas, such as in factories, office buildings, or airports, a multicell, multiuser (MC/MU) system may be the optimum solution. MC/MU systems are a sophisticated means of delivering PCS in defined areas of radio coverage.

In some cases the requirements may even be such that users want PCS in a number of areas, as in a company's headquarters as well as its branch offices, or a factory outside the city center with its sales office in the city center. In those cases several MC/MU systems have to be networked into one, integrated system. Networks like these are of a higher level of complexity and have become available recently.

4.1.1 Network Plus Radio Access

To provide a PCS, one needs a network with the ability to route calls and services to the location where the subscribers actually are, not to the locations where the subscribers are registered. In addition, wherever the subscribers actually are, they should be able to access the same set of services as where they are registered and for which they pay their subscription fees. One way of achieving this service is by connecting databases, which are available at several places in the PCS network, such that subscriber information (like a user's personal number and the set of services he or she has subscribed to) is accessible at any point in the network.

In the network that provides the backbone for GSM, this database connection is accomplished with home location registers (HLR) and visitor location registers (VLR). This chapter's focus is not on these registers but on the radio access part into the network. For simplicity it is assumed that so called *intelligent telecom networks* will have the capability of delivering database services for PCS purposes.

The second building block for PCS is the ability to have (many) users access the network over a radio link. This is the focal point of this paper: radio access to a network provides for local, terminal mobility. Local mobility is defined as the area covered by the MC/MU radio access system. Terminal mobility is defined as the capability for the subscriber to carry an access (voice or data) terminal and access the network for services at any place and any time.

4.1.2 Requirements

The successful candidate for the radio access standard for personal communications has to meet the following requirements:

- *Low-cost terminals:* Personal communications can become truly personalized only if the cost of handsets is low enough to address the consumer market. This requires a technology that leads to low-cost hardware. Uncomplicated terminal design should be possible.
- *Low-cost network infrastructure:* The future of personal communications is envisaged whereby multiple operators offer total or partial area coverage. The required investment in infrastructure should be small, to reduce the operator's outlay.
- *High voice quality:* Personal communications will compete with wired telecommunications. For end users, voice quality comparable to current wired quality is important.
- *Data applications:* The growth of data communications, both in number of users as well as in data-throughput requirements, will also become manifest in personal communications.
- *Features:* Users will demand two types of features. On the one hand, they will request features related to the radio access, such as speech encryption to secure the privacy of their conversations; on the other hand they will require access to network features such as call forwarding, database access, call screening, message waiting indications, and dual-tone multifrequency (DTMF)–related features.
- *Uncoordinated, multioperator situation:* Probably the most important requirement is that numerous PCS operators compete to provide service to end users in the same geographical area (neighborhood, town, or country). This competition sets tough technical requirements on the radio access standard, since it has to support a cost-effective way of handling multiple operators.

4.1.3 The Technology: TDMA

Marketing a PCS based on radio access can be done successfully only if the product is based on a technology that meets the requirements listed in Subsection 4.1.2. The only currently available technology that meets these requirements is TDMA (time-division multiple access), as is applied in DECT (digital European cordless telecommunications).

The principle of TDMA is relatively simple. Traditionally, voice channels have been created by dividing the radio spectrum into (ever narrower) frequency RF carriers (channels), with one conversation occupying one (duplex) channel. This technique is known as FDMA (frequency-division multiple access). TDMA divides the radio carriers into an endlessly repeated sequence of small time slots (channels). Each conversation occupies just one of these time slots. So instead of just one conversation, each radio carrier can carry a number of conversations at once.

The price that has to be paid for splitting up RF carriers into time slots is the bandwidth of each RF carrier. On average, the bandwidth per carrier has to be wider in the case of TDMA than in the case of FDMA. However, the main advantages of TDMA become evident when it is realized that a transceiver, when handling a conversation, is occupied for only a part of the time. In the case of traditional (analog or digital) FDMA systems, a transceiver is fully occupied when handling a conversation.

The fact that in the case of TDMA a transceiver is occupied only for the duration of the time slot creates two important advantages:

- Cost reduction, because one transceiver can handle a number of calls simultaneously;
- Decentralized RF management, because a portable has time available to send and to receive additional information.

TDMA is the access method used in CT3, PHP, and DECT. Table 4.1 summarizes the air-interface characteristics of these systems, in comparison with the older FDMA technology of CT2.

- CT3 is the name of the so-called third generation of cordless telephony.
- PHP (personal handy phone) is the Japanese standard for cordless access that aims especially at residential and public-operated applications.
- DECT is the acronym for the European standard on digital cordless telecommunications. We will focus on DECT.

4.2 DECT

The DECT standard was originally intended to solve the problem of providing cordless telephones in high-density, high-traffic environments, such as offices. It was instigated by the Council of European PTTs (post, telephone, and telegraph service operators) as a European standard for cordless telecommunications, with applications that included residential telephones, telepoint, the cordless PBX, and cordless local area access to the public network.

DECT enables users to make and receive calls when in range of a base station (about 100m in an indoor environment and more than 500m in an outdoor environment). The standard has a seamless handover facility, which allows users to move between base

Table 4.1
Digital Cordless Comparisons

	CT2	CT3	DECT	PHP
Multiple access and duplex	FDMA/TDD	TDMA/TDD	TDMA/TDD	TDMA/TDD
Time frame	2 ms	16 ms	10 ms	5 ms
Speech coding	G.721	G.721	G.721	G.721
RF frequency	864.1–868.1 MHz	862–866 MHz	1880–1900 MHz	1895–1911 MHz
Carrier spacing	100 kHz	1 MHz	1.728 MHz	300 kHz
Channels per carrier	1	8	12	4
Number of carriers (channels)	40 (40)	4 (32)	10 (120)	53 (212)
RF bit stream	72 kbit/s	640 kbit/s	1152 kbit/s	384 kbit/s
Modulation	GFSK	GFSK	GFSK	QPSK
Average power	5 mW	5 mW	10 mW	10 mW

stations during a call without being cut off. This means that users do not notice the range limitations on a base station, because they just roam from base to base. The small cell size does, however, provide advantages for capacity and speech quality. Also, the digital radio link and the 32-kbit/s adaptive differential pulse code modulation (ADPCM) speech code contribute to a speech quality that is as good as wireline quality. The radio link is encrypted to provide absolute call privacy.

DECT uses a multicarrier TDMA/time-division duplex (TDD) format for radio communications between handset and base station. With DECT, 10 radio carriers are available, each 1.728 MHz wide. Each radio carrier is divided in the time domain into 24 time slots, two of which provide a duplex speech channel. When a call is set up, it uses only 2 of the 24 time slots, alternating between transmitting and receiving signals. The remainder of the time can be used by the handset to monitor all other frequencies and time slots and shift the call if a better speech channel is available.

This continuous dynamic channel selection (CDCS) technique is a function under the control of the handset. CDCS is a process whereby the handset is continuously gathering information to decide on selecting better speech channels. This process may occur if, for example, the user moves away from one base station and toward another. The handover is undetectable by the user, which is important in a picocellular environment, where several handovers may be necessary during a short call.

The strength of DECT as a standard for cordless access is not only based on the TDMA principle. The standard has been specified in such a way that it caters for access to a range of host networks, including PBXs, networked PBXs, public switched telephone networks (PSTNs) (including telepoint), GSM networks, packet switched public data networks (PSPDNs), and integrated services digital networks (ISDNs). DECT is highly capable of providing mobile services based on an intelligent host network, and therein lies its strength as a candidate for PCS.

4.2.1 DECT and OSI

The DECT standard has been structured according to the open systems interconnection (OSI) model. The RF access occupies the lower three layers of OSI, but as OSI takes no account of radio transmission uncertainties and handover, DECT has redefined this part into four layers plus a lower-layer management entity (Figure 4.1).

The physical layer (PHL) defines the radio spectrum management. The medium access control (MAC) layer performs three main functions. First, it selects, establishes, maintains, and releases channels. Second, it multiplexes error-control information with higher-layer information into the time-slot packages. Third, the MAC layer provides a reliable point-to-point link. The data link control (DLC) layer provides reliable data links to the network layer. In this way, high levels of data integrity over the radio interface are provided.

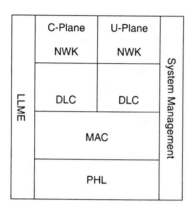

Figure 4.1 DECT reference model.

The fourth layer in DECT is the network (NWK) layer , which is the main signaling layer of the protocol and mainly supports the establishment, maintenance, and release of calls.

Finally, a lower-layer management entity (LLME) has been defined to cater for procedures that concern more than one layer. LLME is intertwined with MAC, DLC, and NWK layers.

4.2.2 PHL: The TDMA Format

DECT's TDMA format is described in Part 2 of the DECT standard, the physical layer. Every radio carrier supports 12 duplex channels, which consist of 12 pairs of time slots per TDMA time frame. Each time frame is available on every carrier. DECT has 10 carriers of 1.728 MHz each. The TDMA time-frame and slot structure for DECT is shown in Figure 4.2.

The TDMA format for the RFP-to-PP connection is time-duplexed with the PP-to-RFP connection. This process is called *time-division duplex* (TDD).

Because DECT uses 10 carriers over which this TDMA/TDD format is applied, it is basically a multiple-carrier (MC) TDMA/TDD technology.

4.2.3 Main Features

The main features of the DECT standard are:

- Dynamic channel selection, a process whereby the portable continuously scans the environment and dynamically selects better channels when they become available;

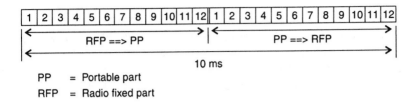

SYNC	A		B		GS
32	48	16	320	4	60
	CTRL	CRC	Information	X	

SYNC = Synchronization
CTRL = MAC layer control data
CRC = Cyclic redundancy check bits
X   = Check bits
GS   = Guard space (equals 52.1µs or 60 bits)

Figure 4.2 TDMA frame and slot structure for DECT (one carrier only).

- Seamless bearer (or channel) handover, the undetectable handover from channel to channel or from cell to cell;
- Two-way call setups, both call originate and call receive;
- Roaming, which is the ability to make or receive calls anywhere in the radio coverage network;
- Authentication and encryption, which provides high-level security.

These features emphasize the unique capabilities DECT as a standard offers.

However, because DECT is specified to provide cordless access to a number of host networks, not all of these features may be available in the applications expected to be on the market in the coming years. For example, a DECT residential set may provide high voice quality and many additional features, but handover is not required in a single-cell environment. Also, roaming in a DECT-telepoint application will be difficult, as long as DECT-telepoints are providing access to the current PSTN.

4.2.4 Idle State

Each RFP in the system is always active on at least one time slot. This can be a traffic slot (i.e., one that is being used for an ongoing call between an RFP and a PP) or a so-called dummy slot.

Both traffic slots and dummy slots contain system and RFP identification. In idle state, a PP scans on a regular base the available time slots and locks to the RFP that has the strongest field strength (and belongs to its own system).

Whenever another valid RFP gives a higher field-strength the PP will lock to the new RFP (although some hysteresis is built in to stabilize the system).

4.2.5 Call Setup

If there is an incoming call for a specific portable, a paging message containing the portable's identification will be sent on the signaling channel of all the active time slots (traffic slots and dummy slots) in the system. This message will be read by the portable. If the portable is locked to a dummy channel, it will send a call request in the corresponding time slot in the second half of the TDMA frame, addressing the RFP to which it is locked.

If the portable is locked to a traffic channel, it will take the best time slot available (with the lowest measured field strength) in the first half of the frame and then send a call request in the corresponding time slot in the second half, addressing the RFP it was locked to in the idle state. (The DECT standard allows for different call-setup procedures as well). After the portable and the RFP control logic have agreed on the time slot to be used, the RE will switch the incoming call to the appropriate RFP and time slot. In case an outgoing call is initiated from the PP, a call request is sent in the same way as for the incoming call. After the portable and the RFP have agreed on the time slot to be used, the RE will allocate a speech channel to the call and send out the dialing information to the network interface.

4.2.6 Handover

The principle of seamless handover can be explained by using a simplified model of a single carrier system with only two cells (RFP1 and RFP2) and two portables (PP1 and PP2). Assume the situation illustrated in Figure 4.3(a). PP1 is in conversation with RFP1 using time slot TS2, a traffic channel. PP2 is in idle state and is locked to RFP2 using time slot TS5, a dummy channel. The dummy channel on TS5 is also monitored by PP1, which is geographically somewhere between RFP1 and RFP2.

Apart from the time slots to which they are locked, PP1 and PP2 will scan all the other 11 available time slots on a regular basis and have information on the status of these alternative time slots stored in memory. When PP1 movies from RFP1 toward RFP2, the field strength measured on TS5 will increase and at a certain moment become higher than the field strength measured on TS2, which will decrease. If that difference is above a certain value, portable PP1 will decide to initiate a handover.

PP1 transmits a handover request on the next possible occasion in TS5 in the second half of the time frame. This handover request includes the identification of PP1 and of

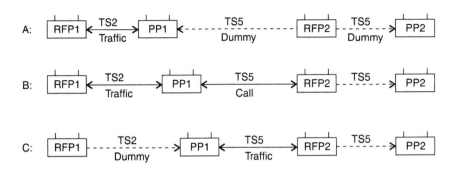

Figure 4.3 A handover example.

RFP2, the base that is addressed. RFP2 will answer in the corresponding time slot TS5 in the first half of (one of) the following frame(s).

After some specific signaling, PP1 and the base station control logic of RFP2 will agree on the time slot to be used, which in this case will be TS5. (Note: If TS5 were a traffic channel instead of a dummy channel, the connection request from PP1 to RFP2 would have been transmitted on the time slot where the lowest field strength was measured by PP1 being the best channel.) During the time needed to set up a new channel to RFP2, PP1 maintains in parallel the existing channel with the ongoing call on TS2 to RFP1 (see Figure 4.3(b)).

When the new channel to RFP2 has been set up, the RE will be informed that a handover will be made, and the RE will then switch the ongoing call to the appropriate base station and time slot. This switching can be done without the user of PP1 noticing it, because the old and the new channels overlap for a short time. After that, the channel to RFP1 will be released (see situation illustrated in Figure 4.3(c)). The handover described is called an *intercell* handover and is the most common type of handover. It makes sure that the system is always in a stable mode with all the portables locked to the nearest base station. No deterioration of the existing call quality will be detected before or after the handover occurs.

4.2.7 CI Conformance Levels

DECT caters for the standardization of a rather wide number of applications, including digital residential telephone sets; public access systems, such as telepoint and radio local loop; wireless PBX (sub)systems; wireless data networks; and a number of other applications.

The core of these DECT products is that the radio interface is as described in the DECT standard. Therefore, the user will be offered the choice of a number of interworking levels. A well-defined level is called *public access profile* (PAP).

PAP is part of the DECT-CI-PROFILE family. This is a level of common interface (CI) conformance, whereby interoperability between equipment of different origins is catered for. DECT-CI also allows for proprietary additions (DECT-CI-PROFILE-PLUS). In the case of PAP-PLUS, a manufacturer could add proprietary additions to the PAP interface. These additions may consist of, for example, nonstandardized features. However, within PAP many features are already standardized.

At a lower level of CI conformance, the DECT-CI-BASE has been defined. At its lowest level, equipment specified to meet the minimum DECT-CI conformance will be using the DECT PHL and some MAC-layer protocols. However, it is also possible to follow the DECT-CI-BASE on MAC, DLC, and NWK layers. Once again, the possibility is created to add proprietary additions.

In this way DECT is a standard that meets four very important requirements:

1. It allows for CI conformity, which will be beneficial both for cost-reduction as for faster market takeoff;
2. It allows for a multiple-operator environment, where a (large) number of uncoordinated systems (which may even address different applications) can coexist in the same physical environment using the same frequency band;
3. It allows for the deployment of vendor-specific (or operator-specific) product-technology additions, creating the opportunity to have product differentiation as well as a future flexible standard;
4. It provides for one and the same product for all of Europe (though differences in software applications are allowed for), creating an opportunity for low-cost volume manufacturing.

4.3 APPLICATIONS WITH DECT

This section describes examples of residential, business, and public access applications of DECT.

4.3.1 Residential Applications

Residential cordless telephones provide the same facilities as fixed residential telephones, but they enable domestic users to move around the house while dialing and when engaged in conversation. They have become widely used in a number of countries, notably the United Kingdom and the United States.

Residential sets are currently available that conform to the CT0, CT1, and CT2 standards. DECT-based products were becoming available in 1993.

It is interesting to note that the manufacturers of DECT-based residential cordless telephone sets have selected DECT as the basis for their products because of the cost advantages DECT brings. This becomes manifest in applications whereby one base station is handling more than one cordless telephone. The time multiplexing of calls, inherent in the DECT standard, offers an immediate cost advantage.

4.3.2 Cordless PBX

The primary aim of the DECT standard was to meet the need for cordless extensions on large PBX systems. The office worker uses a pocket-sized portable telephone, which provides all the facilities of a wired extension to the office PBX, wherever the user is on the premises. Every handset has its own unique identity, and its location is tracked by the PBX. The identity allows the handset to be called when an incoming call is directed to it.

The cordless PBX eliminates one familiar problem: telephoning someone who, while in the building, is not at his or her desk to receive the call. It is ideal for staff who by the nature of their jobs are difficult to locate, such as messengers, maintenance staff, and warehouse staff. The cordless PBX also reduces the amount of telephone wiring needed in offices. This makes office reorganizations easier and reduces the administrative workload. For example, when an employee moves to another office, his or her extension number does not have to be changed, nor does the telephone system need to be reprogrammed.

4.3.3 Telepoint or Public Access

Telepoint systems enable the subscriber who is in range of a base station either to make or to make and receive telephone calls. A telepoint installation that allows only outgoing calls can be compared to a public telephone box, but with fewer restrictions: the user is not confined to the call box when making a call and does not have to worry about having suitable coins or a credit card available. In theory, base stations will also be more widely available than public call boxes.

4.3.4 Radio in the Local Loop

Another potential application of cordless technology is *radio in the local loop*, or RLL. This application uses radio to make the final link between residential subscribers and the PSTN. This can have advantages for both the local network operator and the residential subscriber.

Hard-wired connections in the local loop are both expensive and difficult for network operators to install and maintain, whether provided as overhead or underground cables. To install a new connection, engineers need access to the subscriber's premises, and maintenance may result in a lack of service for a period of time, both of which are inconvenient for the subscriber.

The huge investment needed for local cabling is also one of the largest obstacles to truly competitive service provision in a liberalized telecommunications environment.

Competitive services offer cost advantages to the residential subscriber and business opportunities to potential new network operators. No market forecasts for the introduction of cordless local loop services have been made. The introduction of services will depend on the network operators rather than on market forces.

4.3.5 Personal Telephony: The Cordless Concept

Personal telephony involves three basic system requirements: a personal handset, a radio transmitter to communicate with user handsets, and a network with the intelligence to follow a handset wherever it goes.

As DECT is deployed in the coming years for residential sets, business cordless, telepoint, and RLL, we can expect the availability of many handsets and many radio transmitters in the market. This will definitely facilitate the startup of DECT-based personal communications.

However, to complete the concept of PCS, one needs a network with certain intelligent capabilities. These networks do exist. One of them, interestingly, is the GSM network (not to be confused with the GSM access part). Another option is to add databases to a digital network. This is happening, for example, in the world of private digital networks (PBX networks).

4.3.6 Market Availability

Because of the flexible nature of the DECT standard, that is, the fact that it can be applied for various cordless access purposes, we should expect manufacturers to focus on various applications.

Product availability was expected as of 1993, and some manufacturers have already launched DECT products. Olivetti/Sixtel's primary focus is on a wireless local area network (LAN). Siemens has started marketing a residential set (SC/MU) with intercom facilities. Ericsson is marketing a PBX add-on (MC/MU) system that creates a business PCS environment in, for example, hospitals, factories, and offices. Virtually all manufacturers announced products for availability in 1993 and 1994, ranging from low-cost consumer products to sophisticated PCS access products.

All these products comply to the DECT standard as completed and ratified in 1992. With type-approval common technical regulations (CTRs) 6 and 10 finalized by mid-1993, nothing is stopping DECT from becoming a successful example of market-oriented standardization.

4.4 PCS IN NORTH AMERICA

Whereas in Europe the focus has been on standardization of technologies (in particular, GSM and DECT), the United States has adopted a different approach.

The Federal Communications Commission (FCC) has allocated a range of spectrum for licensed and unlicensed PCS. Two bands are allocated for major trading area (MTA) PCS services: the A-band (1850–1865 MHz and 1930–1945 MHz) and the B-band (1865–1880 MHz and 1945–1960 MHz). For basic trading areas (BTAs), of which there are 492 in the United States, 5 bands (C-G) are identified. The C-band is two times 10 MHz (1880-1890 MHz and 1960–1970 MHz), and the D, E, F, and G bands are contiguous two times 5-MHz bands (2130–2150 MHz and 2180–2200 MHz). For unlicensed PCS, 40 MHz is allocated, for both voice and data.

At the time of writing, the FCC Notice of Proposed Rule Making (NPRM) was still being debated, and it is therefore difficult to draw conclusions. However, the nature of the PCS allocations, both for licensed as well as for unlicensed services, seems to allow for DECT or DECT-like technologies to be deployed.

In 1993, Ericsson and US WEST started a field trial for PCS, based on DECT cordless access products. The trial was scheduled to start toward the end of 1993 in Boise, Idaho. Initial results indicate that DECT is a likely candidate for providing local mobility and PCS.

4.5 CONCLUSIONS

The revolution that cellular communications has aroused will find its logical sequence in achieving a world of personal communications. In this respect, radio technologies play an important role. However, the challenge in current radio communications is to offer high-density solutions without compromising voice quality.

This is exactly what has been achieved with the DECT standard. The standard offers ways to provide cordless access to host networks for a wide range of applications. DECT will prove to be as important to achieve the goal of PCS as cellular has been.

SELECTED BIBLIOGRAPHY

Åkerberg, D., et al., "A Business Cordless PABX Telephone System," *IEEE Communications Magazine*, Vol. 29, No. 1, Jan. 1991.

ETSI, *European Telecommunication Standard, DECT, ETS 300* 175, Parts 1-9, and *ETS 300 176*.

ETSI/TC-RES, *A Guide to DECT Features*, RES(92)09 Annex 4.

Lauridsen, O. M., "Personal Communications," Presentation Paper, 35th RACE Concertation Meeting; Telelaboratoriet, Telecom Denmark.

Hoek, H. B. van der, "DECT," *European Communications*, 1993.

Trivett, D., *DECT*, Datapro Publication, McGraw-Hill.

Beek, H. van, "Business Cordless and DECT: Worldwide Experiences," Presentation Paper, 2nd Annual Wireless Information Networks Conference; Ericsson Business Mobile Networks BV.

Mulder, R. J., "DECT, A Universal Cordless Access System," *Philips Telecommunication Review*, Vol. 49, No. 3, Sept. 1991.

Chapter 5

PCN Service and Its Implementation Using DCS1800

Peter Ramsdale and Robin Potter

Personal communications is not defined by a specific technology; rather, it is described by the features a user would wish for from an individual telecommunications service. Although a single definition is not possible, the features sought by most people can be encompassed by a common vision. The objective of a personal communications network (PCN) is to meet this vision as fully as possible and bring the mobile phone to the mass market.

Mercury One-2-One launched its service in the United Kingdom in 1993, and can thus claim to be the first operator to establish a commercial PCN service. The initial coverage was of Greater London (within its orbital motorway, the M25), but it will extend over the whole of the country by around the turn of the century. The system implementation is based on DCS1800, a variant of the GSM standard providing for operation in the 1800-MHz band. In this chapter we see how the features defined for the DCS1800 standard have enabled Mercury One-2-One to realize a service that embodies its perception of the personal communications vision.

5.1 MARKETING DRIVE

Market research reveals a consistent set of features that form a vision of personal communications recognized by the majority as representing their ideal individual communication system. However, any practical technology will favor some attributes at the expense of others. Thus, PCN sets out to satisfy the most important aspects from a low-cost base. In addition, where compromises have to be made, solutions that can be improved over time from predictable technology advances such as better semiconductor devices are sought. For example, it is considered more acceptable for early handsets to be slightly larger or to provide less than the desired battery life rather than for the network never to be able to offer contiguous coverage.

PCN is essentially a single service that replaces mobile, cordless, and fixed phones with individual pocket or handportable phones. The emphasis is very much on the handportable and not the mobile car phone, although phones are expected to work if the user is in a vehicle, and car adaptors are an available option.

The network provides high-quality speech, ideally as good as that of the fixed public switched telephone network (PSTN). This requires good-quality voice coding with little degradation from the mobile network. Thus, PCN radio coverage must be very good, with its service area covered contiguously with an adequate radio signal strength. The radio coverage supports handsets both outdoors and in buildings. The geographic coverage is targeted on built-up environs and regularly visited areas such that there is a simple marketable proposition to the customer that the phone will work within a clearly specified area. "Phones for people, not places" and "The mobile phone for everyday, for everyone" are more than just useful slogans; they are the keystone of the PCN philosophy, which is to treat people as individuals by aligning the telecommunications network to suit their specific needs.

PCN aims to create a mass market for mobile phones. Any restrictions or complications in the ways in which the phones can be used will limit its take-up. This means that the handset should always be able to make or receive calls anywhere within its service area, with its basic method of use identical to that of a conventional fixed phone. In addition, calls once established should be maintained whether the user is stationary or mobile. Other features provided as enhancements to these basic functions can include alerting lights or tones associated with cheap tariff zones and call waiting. A particularly attractive service is an integral voice-mail within the network, which provides two important features:

- There are times when calls to a personal phone can be an intrusion; for example, during important meetings or the theater. If the call is diverted to a voice-mail system and an alerting light set off on the handset, the message can be accessed at a more convenient time.
- If the PCN phone is outside its geographic coverage area, incoming calls can still be completed by the voice-mail system. When the phone returns to the radio coverage area, a message waiting light is activated on the handset, and the voice-mail box can be accessed.

The benefits of a fully mobile pocket phone are clear, but for widespread acceptance the cost of usage must be low. The key is to put in place a low-cost infrastructure using economies of scale to aid its implementation and high levels of usage to spread fixed costs over many calls. Unlike conventional mobile cellular radio, the costs of interconnection to the PSTN for completing the delivery of calls to and from fixed telephones, are negotiated to be similar to fixed telephone rates, and use local and trunk bands. In contrast, previous implementations of mobile cellular have treated all calls as long distance trunk calls with high interconnect rates, which lead to high tariffs for the service.

The handset is the customer's direct contact with the PCN, and its attractiveness has a major effect on the potential purchase of the service. Size, weight, and ergonomic appearance must all be considered. The physical length of any telephone is governed by the distance between the mouth and the ear, although flip-phones can fold to a shorter length for carrying. However, the overall weight and volume are determined principally by the size of the battery. With today's technology for digital phones, the batteries required for a full day's usage, which consists of intermittently making and receiving calls and continuously remaining affiliated to the network, are larger than ideal. As the power consumption of semiconductors continues to fall with succeeding generations of submicron feature sizes and higher-capacity (nickel hydride) batteries become widely available, handportables will reach the point where further reductions in size and weight are of diminishing value.

5.2 PCN STANDARD REQUIREMENTS

To meet the functionality required by the vision of personal communications, the only suitable technology basis is that of cellular radio. Fixed networks can provide personal mobility, such that users can use any access point, but some form of radio access is needed for terminals to be portable. Cordless technologies can be used to provide radio links to access points on a fixed network. Examples of cordless telephones with defined air-interface access are CT2 and DECT (see Chapter 4), but these standards do not define the routing, switching, and service functions that are provided by the service network.

However, cellular network standards define complete telecommunications networks, including full terminal mobility, such that handsets are continuously affiliated with the network, and handover takes place from cell to cell such that handsets maintain calls even when highly mobile.

A further advantage of using cellular rather than cordless technology for PCN is in the range of cell sizes. Cordless access technologies have been designed only for quite small cells (up to 200m) and "stretching" the cells is difficult. In the case of CT2, range is limited by fast fading because there is no error protection on the speech channel. For DECT, the data rate (bit period = 868 ns) was designed primarily for indoor operation, where delay spreads are short (typically 50 ns), and DECT does not have the necessary channel equalizer for greater delays. Cellular air-interfaces tend to be more complex and require more elaborate transceivers, but these are suitable for a wide range of cell sizes, indoors and outdoors, such that they can be used in small (high-capacity) cells and large (wide coverage area) cells.

The U.K. Department of Trade and Industry (DTI) required PCN to be based on existing standards produced by the European Telecommunications Standards Institute (ETSI), because doing so would enable early introduction of the service and encourage harmonization within Europe with the consequential advantage of market-size economies (see Chapter 2). In their PCN license applications to the DTI, the successful applicants proposed adapting the 900-MHz GSM standard for operation in the 1.8-GHz band (DCS1800). GSM was chosen because it provides a good match to PCN requirements, defining a complete digital cellular radio system incorporating these elements:

- Interfaces for radio, transmission, and switching networks;
- Network, radio link management, and location facilities;
- Large set of telecommunication services (aligned to the integrated services digital network, or ISDN);
- Support of cell sizes giving economical wide area coverage;
- Support of high-speed mobiles, including handover;
- Potential for use in the local loop;
- Defined interconnection to the PSTN.

At the same time, it was recognized that enhancements to the GSM standard were needed not only to translate its operation to the 1.8-GHz band but to refine the standard for:

- Support of small, low-power handsets;
- Close-proximity working;
- Wider available frequency band (150 MHz compared with 50 MHz for GSM);
- Improved internetwork roaming capabilities.

One enhancement has been to support a multioperator environment by incorporating *national roaming*. *International roaming*, as defined by GSM, caters for mobiles belonging to a network visiting another country and receiving service from a network of that country. National roaming allows operators of an individual country to restrict roaming to selected areas of the visited network. The key feature of the standard is to prevent a mobile from remaining on the visited network when the home network is available.

5.3 DCS1800 RF ASPECTS

Before DCS1800 could be developed, a frequency band of operation had to be established. The band 1710 to 1880 MHz was ratified as the allocation for PCN in Europe by CEPT and consists of two bands of 75 MHz each with a 20-MHz separation for the split duplex operation, providing a maximum theoretical capacity of 375 radio carriers, each with 8 or 16 (half-rate) voice/data channels. It is three times the allocation to GSM in the 900-MHz band and is consistent with supporting the peak traffic densities anticipated for PCN. Initial allocations from this band have been made for DCS1800 systems in the United Kingdom and Germany.

The RF performance requirements of a radio link are complex and interrelated. To arrive at a self-consistent set of specifications that meet system needs and can be implemented at a reasonable cost, a rigorous and methodical approach was needed. To determine the system requirements, a range of scenarios was formulated, defining the relative physical positions of mobile and base stations likely to be encountered in operational conditions. Other constraints such as the channel spacing, frequency reuse assumptions, and the band of operation were also included.

Six scenarios were examined:

1. Single mobile station (MS), single base transceiver station (BTS);
2. Multiple MS and BTS where operation of BTSs is coordinated (single operator);
3. Multiple MS and BTS where operation of BTSs is uncoordinated (multiple operators);
4. MS in proximity to another MS;
5. BTS in proximity to another BTS;
6. Proximity to other, non-DCS1800 systems.

Limit conditions were selected that were likely to occur relatively frequently in PCN usage. The impact of each scenario on the radio performance requirements was assessed to identify the most demanding scenarios. This enabled the requirements to be determined from a purely system point of view. These were considered alongside the complexity and cost of implementation, and, where necessary, compromise specifications were reached.

The two issues that have the greatest impact on the RF specifications and their practical implementation are the power output of the MS and the requirements for close-proximity working. These issues determine the majority of RF performance requirements, including blocking, spurious emissions, and intermodulation characteristics.

5.3.1 Mobile Power Class

The selection of the power class is a tradeoff between the requirements for small, light-weight handsets with good battery life, high-quality RF coverage in a range of environments, and the need to minimize the cost of the radio network. Increasing the power improves the range but has a variety of implications. It increases the size of the handset and may lead to greater problems in controlling wideband noise and spurious emissions, and the output RF filter needs a higher power-handling capability.

On the other hand, lowering the power means that more base stations will be required, and so the cost of the radio network increases. However, adding more base stations means that the network capacity is increased, which is in line with the idea of PCN as a mass-market service. After study of various options and following discussions between manufacturers and operators, the MS power classes were defined as 250 mW and 1W.

Apart from these considerations, the choice of low-power classes only for PCN was also taken to aid close-proximity working. Since PCN intends to use small cells and be a high-capacity service, it is likely that there will be MSs from another operator close to a BTS on full power. This places stringent demands on the MS and BTS performance, but these problems are alleviated to an extent by the use of a low-power class.

5.3.2 Close-Proximity Operation

In a high-capacity PCN, not only will mobiles be operating close together, but as a result of small radio cells (for capacity) and low base station height (to control interference), there is a high probability that mobiles will operate close to base stations. This is not so significant in a single-operator environment (scenario 2) since the mobile will be powered down as it approaches the base station. However, in a multioperator environment (scenario 3), as will occur in the United Kingdom, mobiles can potentially be on full power and close to a nonserving base station (the ''near-far'' problem). This places stringent requirements on mobile and base station RF performance, particularly in respect to blocking to avoid destination, spectra due to modulation and switching of the TDMA burst structure, and noise and spurious emissions. A mobile near limit range would be transmitting at full power. This mobile may be close to a base station of a second network, which is receiving signals close to limit sensitivity from its mobiles. In this case, out-of-band emissions from these mobiles can desensitize the base station line receiver of the first network, thus reducing its range. The RF specification must protect adjacent bands of different operators, or the base station range may be limited and capacity reduced through excessive interference.

Similar issues arise with the collocation scenarios (4, 5, and 6). To determine the worst-case coupling between MS and BTS, a variety of physical conditions were considered, taking into account antenna height, gain, and MS-BTS separation. A figure of 65 dB for the worst-case MS-BTS coupling was chosen (given, for example, by an MS on boresight of a 10-dBi base station at a distance of 30m) and used to derive the RF performance requirements.

5.4 DCS1800 RADIO INTERFACE

The radio interface, or air-interface, of GSM/DCS1800 uses TDMA to provide 8 or 16 channels/carrier with a gross data rate of 22.8 kbit/s in each full-rate (8 channels/carrier) channel and a frame periodicity of 4.6 ms.

To accommodate the needs of the radio channel and the various signaling and control requirements, a complex TDMA frame structure has evolved (Figure 5.1). A burst of data is transmitted during each active time slot of the 8-channel TDMA frame. Each time slot transmits 116 encryped message bits while in the center of the time slot; 26 bits are used as a training preamble sequence for the equalizer of the receiver to create a ''model'' of the radio channel and counteract the effects of multipath time dispersion. At the end of each time slot, a guard period of 8.25 bits is provided to allow for uncertainties in the arrival time of TDMA time slots at the base station from mobiles at varying distances. Such a small guard period is made possible by the use of a timing advance control whereby a base station continually monitors a mobile and instructs it to advance or retard its transmit timing so that time slots arrive at the base station at an approximately correct time.

TDMA frame

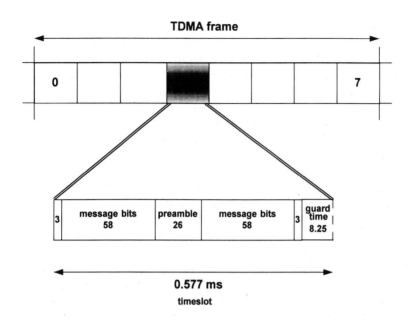

Figure 5.1 TDMA frame structure.

Control information is generally mapped onto time slot 0 within the TDMA frame and a 51-frame multiframe is created to further multiplex control information channels. Frequency correction and synchronization data are delivered within the time slot 0 structure at periodic intervals.

Traffic channels may be organized as full rate (eight per TDMA frame) or half rate, the latter being generated by using every other frame for information pertinent to the given traffic channel. It is possible to mix both full-rate and half-rate channels within a frame, although clearly a given time slot can be used only in full-rate or half-rate mode at any particular time.

The full-rate speech coder operates at a net rate of 13 kbit/s. Speech data are then heavily protected against errors by channel coding that takes the gross data rate up to the 22.8 kbit/s of the full-rate channel. The channel coding takes account of bit significance in the speech coder, and interleaving of the data across eight TDMA frames is also applied to minimize the effect of short error bursts on the radio channel.

The TDMA structure is also used by the mobile to decode signals from a number of surrounding base stations during a call (this is done in a sequential fashion) and report back to its current home base station the signal-quality parameters it has measured. This information can then be used to provide a basis for computing handover options—measurement by the mobile providing a better up-to-date view of the radio environment than a sequential polling of alternative base stations.

The instantaneous data rate over the radio channel is 270 kbit/s. Gaussian minimum shift keying (GMSK) modulation is employed with a normalized bandwidth of 0.3, enabling a channel spacing of 200 kHz to be used. At this data rate, multipath channel equalization is required, and an extensive measurement campaign has identified that equalization of up to 16 ms of multipath delay is adequate for most practical cellular-PCN environments. (Note: It is clear that in mountainous environments delayed reflections of greater than 16 ms can be encountered; however, cellular engineering can, in general, eliminate or at least minimize the effect of such reflections. Once outside the equalizer range, it is necessary that the level of such an unwanted signal be kept well below the carrier- and interference-handling capability of the system.)

Frequency hopping is an optional network capability in DCS1800 (all mobiles are implemented to support the hopping capability). Hopping occurs at the TDMA frame rate, that is, around 217 hops/s with the hop sequence being communicated to the mobile at call setup and handover times. At each base station (or sector of a sectorized base station), one carrier supporting the broadcast control channel (in time slot 0) does not hop so that mobiles can always listen for commands. Frequency hopping provides an ability to further counteract multipath fading over and above that already achieved with channel coding and interleaving and antenna spatial diversity (which is generally provided only at the base station). In addition, frequency hopping provides for a better statistical distribution of interference, and it is anticipated that its use will enable efficient frequency reuse within the cellular radio network.

5.5 DCS1800 NETWORK INTERFACES

The architecture of a GSM/DCS1800 network is illustrated in Figure 5.2. The figure shows a set of subsystems and interfaces that are defined in detail in a series of ETSI recommendations [1]. The key features of these recommendations are the description of open interfaces to the OSI framework and the use of ISDN standards for signaling and network functions. The interface approach gives manufacturers flexibility of implementation while allowing operators flexibility in equipment procurement.

The base station controller (BSC) controls and manages a number of base transceiver stations (BTSs) by providing the lower-level control of cellular functionality. ISDN LAPD signaling protocols are used within this base station subsystem and over-the-air interface.

The mobile services switching center (MSC) is primarily concerned with routing calls to and from mobile stations. The home location register (HLR) contains the customer information required for call routing and administration, including class-of-service data identifying which services a particular customer is allowed access. Associated with each MSC is a visitor location register (VLR), which stores information, including detailed location data, about all mobiles currently active within that MSC's area of control.

Network access is controlled by algorithms that carry out rigorous authentication of the mobile stations. The equipment identity register (EIR) provides an up-to-date check

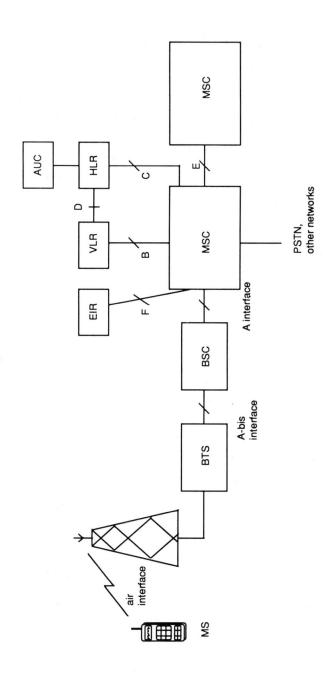

Figure 5.2 DCS1800 network interfaces.

on the validity of mobile equipment, while the authentication center (AuC) checks the validity of the subscriber identity module (SIM). A feature of GSM/DCS1800 is that access to the network is granted through a SIM, which is generally mounted on a smart card plugged into the mobile station to personalize the equipment to a particular customer. Security is provided by the use of challenge/response pairs of signals for authenticating access to the network and encryption keys for decoding enciphered speech, data, and other signaling passing over the air-interface. All speech is digitally encoded at 13 kb/s (full-rate codec). A cryptographic algorithm, A5, produces ciphertext out of cleartext using a common cipher key (Kc). Kc is produced by an algorithm, A8, based on mutual agreement between the mobile station and the fixed part of the system. An algorithm, A3, produces a signed response to a challenge to authenticate that the user is a valid subscriber. The algorithms A3 and A8 are contained in the SIM that holds security and other subscriber-related information.

Network signaling between major elements makes extensive use of the CCITT common channel signaling system no. 7 (C7). The extension to C7 for mobile networks is known as the mobile application part (MAP). MAP supports communication between the MSC, HLR, VLR, and EIR providing functionality such as location, updating of the HLR and VLR, inter-MSC handover, and authentication.

5.6 DCS1800 INFRASTRUCTURE SHARING

One of the differences between GSM and DCS1800 is that the smaller cell sizes of an 1800-MHz network could make the service financially unviable in outlying areas. Therefore, a new roaming approach has been developed to reduce the cost of the network.

5.6.1 National Roaming

A key enhancement of the specification of DCS1800 has been the incorporation of national roaming. *Roaming* refers to a mobile of one network receiving service from another. National roaming allows operators to restrict roaming in a single country to specific areas. This service is additional to the normal international roaming capability of GSM, which caters for mobiles belonging to one network visiting another country and receiving service from a network of that country.

National roaming was designed to deal primarily with the situation where multiple PCN networks are being deployed in a country but individually do not provide nationwide coverage. The aim is for mobiles to automatically switch between networks according to availability of coverage but return to their own (home) network when coverage from it becomes available. This gives the user a wider coverage area and reduces the cost of the networks by sharing between operators. The manner in which this is achieved is as follows.

The network is divided into *location areas* to allow the network to track and page mobiles efficiently. The size of the location area is set by the operator, and this feature provides a useful mechanism for implementing national roaming. Figure 5.3 shows the

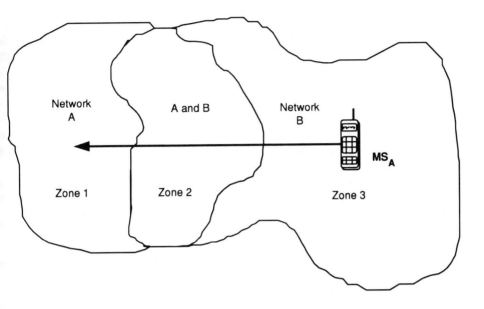

Figure 5.3 National roaming.

areas of coverage of networks A and B whose operators have reached a national roaming agreement. Three zones are defined by the respective location areas of each network. A mobile station belonging to network A (MSA) starts a journey in zone 3. It is located in a location area in zone 3 and receives service from network B. When MSA moves into a location area of network A in zone 2, where coverage is available from networks A and B, the request by MSA to network B to update the location area is rejected, causing the mobile to search for and to access its home network. This is the key step; without the national roaming enhancements, the mobile would remain on the visited network B until it lost coverage from it, that is, until it reached the boundary between zones 1 and 2. While MSA remains in zones 1 and 2, it continues to receive service from its home network. These operations take place without intervention from the user and extend the coverage for MSA by the area of zone 3.

This type of operation reflects one of the underlying principles of GSM/DCS1800 that, wherever possible, a mobile should receive service from its home network, that is, the one subscribed to by the user. In areas where coverage is available from more than one network, the aim is to minimize the time that a mobile is not served by the home network and as part of the work of the phase 2 and phase 2+ standards programs, new procedures are being defined to achieve that aim optimally.

5.7 SHORT-MESSAGE SERVICE

The DCS1800 standard provides a structure for delivering a full range of telecommunications voice and data services using modern signaling and control structures. In addition to this, the short-message service (SMS) is a feature that provides for delivery of messages of up to 160 characters both to and from the mobile in a connectionless manner (that is no speech path setup is required). SMS may be delivered both to addressed mobiles (point to-point service) or on a general broadcast basis from individual or groups of base station (cell broadcast mode). This latter mode is particularly useful for general or localized information services.

In the PCN environment, messaging (both voice and data) can provide a powerful complement to the high-quality voice mobile service. SMS functionality, linked to voice messaging systems, opens up a new vista of service opportunities and will be a major feature of the DCS1800 service offerings in the future. One simple example is the delivery of a voice message waiting signal to the mobile, which is sent when the mobile reactivates into the network, indicating without intrusive interruption that a message has been left while the mobile was unavailable. Such a feature—and there are many variations of voice and data messaging that can be exploited—begins to put into customers' hands a telecommunications product over which they can exert control and yet be given the assurance of being contactable.

5.8 IMPLEMENTATION OF PCN

Although the elements of a mobile cellular network and PCN are similar, the ways in which the networks are implemented can vary significantly in order to create their different market objectives. Traditionally cellular networks were initially deployed with as few cells as possible to provide coverage just over areas frequently visited by mobiles, for example, city centers and motorway corridors. Cellular tariffing has little call-distance structure and comparatively high charges, and it is positioned as a premium service ideal for those requiring wide-area mobility. As the business develops, enhancement of coverage quality and capacity can be financed by the revenue from the existing customers. The better coverage increases suitability for handportable users. Eventually the network can be used for other types of customer, for example, someone desiring occasional emergency service can be offered a package with lower ownership charges and very high usage charges, because the network build costs have already been paid for by the principal users.

PCN operators seek to offer high-quality communications with a customer base requiring contiguous coverage over the community of interest where the service is being sold. In general this community is at least a city plus commuter environs covering hundreds (possibly thousands) of square kilometers. PCN requires a very large initial investment to ensure that the network quality meets the marketing requirement from launch. By building the "final" network at the outset, the total costs are less due to economies of scale in manufacturing and deployment and also due to avoiding the need for expensive upgrading, cell splitting, cell replacement, and replanning commonly experienced in evolving cellular networks. However, because none of the initial network can be financed by

evenue from users, PCN requires long-term investment commitment from its backers
nd recognition of the considerable period before profitability.

The tariffing of PCN must be suitable for a mass market. Although a premium is
ossible for its mobility advantages over the fixed phone, when used near the user's home
he increment over PSTN rates should be small while local and trunk rates should be
upported. This requires interconnection between the PCN and the PSTN at sufficient
oints to access the different tariff bands. Interconnect rates have to be agreed on by the
CN and PSTN operators such that they can cover their costs for their parts of delivering
alls across the two networks. By deciding to make only a small profit margin per call
r by offering innovative tarif packages such as free off-peak local calls, PCN rates can
ncourage high usage and achieve profitability from volume.

DCS1800 has provided telecommunication operators with an opportunity for the
nitial step into the PCN marketplace:

1. The core network and mobile elements were based on the GSM 900 technology,
   which was in production and entering public service.
2. Radio technology to implement the 1800-MHz operating frequency was available.
3. The DCS1800 standard was optimized around the handportable product with both
   low-power portables and base station technology.
4. The characteristics of the 1800-MHz operating band results in a small cell structure
   that is compatible with the PCN concept.
5. The 1800-MHz band is, in general, occupied by fixed radio links for which alternative
   technologies exist, and clearance of the band could be more readily effected than
   attempting to manage coexistence and transition between first- and second-generation cellular systems at around 800 MHz to 900 MHz.

The DCS1800 standard represents the technical foundation for the implementation of the
network, and as such it provides for a wide range of options and design variants suitable
for PCN. However, the specific network design and implementation are equally important
in determining the quality and type of service actually offered to customers. A comparison
of the different implementation provisions between "traditional" mobile cellular radio
and PCN is shown in Table 5.1.

5.9 PCN RADIO NETWORK DESIGN

The initial implementation of PCN is based on provision of a high-quality small-cell
network (cell radius from less than 1 km in a dense urban environment to 5 km in a rural
environment). Radio coverage and system parameters are optimized for the low-power
handportable, and emphasis is on generally providing a significantly higher statistical call
success and quality level for the handportable than current cellular networks can provide.
Coverage targets are set for both indoor and outdoor usage.

The radio cellular design requires the establishment of a suitable radio link path
budget; a typical calculation for DCS1800 is set out in Table 5.2. A 1W (peak) handportable

Table 5.1

Comparison of Traditional Cellular and PCN Implementation

Traditional Cellular	PCN
Optimized to mobile	"Phones for people"
	Mass market economies of scale
	—Handportables
	—Network build
	Greater self-provision of links
	Common handset specification
	—Prioritizing market offering
	Advanced but simple-to-use features
Wide-Area outdoor service	In-building and outdoor coverage
	Microcellular techniques for high capacity
	Combination of cordless and cellular attributes
Trunk PSTN interconnection	PSTN interconnection aligned with coverage

Table 5.2

Typical DCS1800 Radio Link Budget

Mobile peak output power (1W)	+30 dBm
Effective mobile antenna gain	−3 dBi
BTS antenna gain (sector)	+17 dBi
BTS feeder loss	−2 dB
− (BTS receiver sensitivity)	104 dBm
Overall (nonfading) path loss	146 dB

is assumed, and the "effective" mobile antenna gain reflects the relatively low gain and efficiency obtainable from the antennas of small handportable transceivers.

The radio network designer depends on the diversity capability of the DCS1800 system with its combination of time and frequency diversity within the scope of the standard and the effectiveness of BTS antenna spatial diversity to cope with the fading environment that the mobile experiences.

In the dense urban environment, cell implementation is best effected using traditional sectorized structures typically using 120-degree sector configuration; this generally provides for optimal radio coverage with a high-gain directional sector antenna. Spatial diversity on the BTS receive antenna is necessary to provide a balance between the higher-power BTS downlink and the mobile uplink. Frequency efficiency is also optimized with the sectorized structure. DCS1800 has a carrier-to-interface ratio (C/I) of around 12 dB for a good-quality speech, significantly better than existing first-generation analog systems. As a result, more efficient frequency reuse structures can be implemented. A theoretical cell frequency repeat pattern approaching four is within this C/I capability, and with the application of frequency hopping even tighter reuse may be contemplated. It is likely, however, that the practical issues of BTS location and interference propagation characteristics are such that the deployment of such tight reuse arrangements will be impossible in the PCN small urban cell environment. Microcell arrangements can, however, be used, creating exceptions to frequency reuse structures with both base station and mobile operating at lower than normal power levels.

Microcells are small cells whose base station antenna is below rooftop height, so that the RF coverage is confined to a small area. The next stage of evolution may include such microcell structures for coverage and capacity enhancement into buildings where large numbers of people gather, such as airport terminals, railway stations, and shopping malls. A further development would then be the exploitation of "private" cells within offices to provide internal business communications.

Ubiquitous deployment of microcells in a PCN environment requires a very fast handover-processing capability, which was not initially available on DCS1800. However, it is practicable within the first phase of the standard to exploit isolated microcells within the general macrocell environment by careful attention to handover parameter setting, that is, only allowing calls to originate within the microcell and using the overlaying macrocell as the single target handover candidate.

High-quality coverage and grade of service demand advanced, efficient radio network techniques as well as a sufficient spectrum allocation. The air-interface is in effect a traffic concentrator where blocking can occur just as in other parts of the fixed network. Layered cell hierarchies in which a layer of macrocells is overlaid by a layer of microcells located at traffic peaks can offer substantial increases in traffic capacity [2]. This approach makes efficient use of the scarce radio spectrum and, when allied to the 2×75-MHz potentially available to PCN, will provide for very high capacity systems.

The microcell can be used to increase the capacity of the network because it permits greater frequency reuse and can provide many channels in a small area. However, for contiguous coverage a layer of macrocells is also required. Before microcells can be implemented, they need the development of small base stations and new design techniques, which require advances in technology and enhancements to the standards. The classic problem is the risk of dropping calls from high-speed mobiles when they leave the microcell's coverage, for example, when they turn a corner and become shadowed by a building. An enhancement has been developed for Phase 2 of the DCS1800 standard whereby stationary and slow-moving mobiles are encouraged to access the microcell, while ensuring that fast-moving mobiles remain served by macrocells, as shown in Figure 5.4.

A timer mechanism is implemented in the mobile such that when the mobile first receives radio coverage from the microcell, the effective serving cell area is deliberately minimized. If the mobile is still within the microcell coverage area after a penalty time of, for example, 2 minutes, the serving area is increased. In the case of the high-speed mobile, it has already left the coverage area of the microcell and therefore continues to

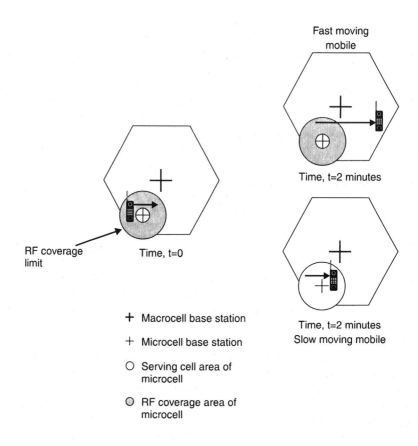

Fast moving
mobile

Time, t=2 minutes

RF coverage
limit

Time, t=0

+ Macrocell base station

+ Microcell base station

○ Serving cell area of
microcell

◑ RF coverage area of
microcell

Time, t=2 minutes
Slow moving mobile

Figure 5.4 Microcell handover.

be served by the contiguous macrocell coverage layer. However, the slow-speed mobile is still within the microcell coverage area and it selects the microcell for its service. This simple mechanism distinguishes the majority of high-speed from slow-speed mobiles.

Outside the urban environment, the same pressures on capacity and frequency efficiency do not apply. In addition, environmental considerations relating to antennas and towers are increasingly a consideration for planners. An omnidirectional antenna arrangement is likely to provide the most effective solution, and technology is now able to provide relatively high-gain (12 dBi) omnidirectional antennas with good beam patterns. Lightweight towers can be used in "green field" situations, and a solution providing excellent radio performance may now be combined with an environmentally acceptable implementation, an example of which is shown in Figure 5.5. Two antennas are used for receiving diversity with a duplexer to permit one antenna to also serve as a transmitting antenna. The spacing of around 5m gives adequate transmit-receive isolation and sufficient separation for spatial diversity in the multipath conditions of rural areas.

Figure 5.5 Typical antenna arrangement.

5.10 PCN TRANSMISSION NETWORK

A transmission network is required to link radio cells to the BSCs and MSCs. The network should provide a sufficiently high availability to achieve a high-quality network. On a link-by-link basis, the required availability level differs since it is clearly less important to lose a single cell for a short period than a considerable portion of the network. The simplest method of providing connectivity is a tree-and-branch structure, but this leads to the network being very susceptible to failure of the higher multiplexed links. Resiliency can be improved by hot-standby parallel paths or ring structures, although these must be of greater capacity and include equipment capable of rerouting.

If microwave links are designed to have an availability of 99.99%, then on average all links will be unavailable for 1 minute per week. The unavailability is primarily due to propagation outage, which in reality means there will be a 5- to 10-minute outage about six times a year spread over a 4-month yearly cycle (e.g., wet snow in January and February and thunderstorms in July and August) with a few random events at unpredictable times.

With its proportionately greater number of cells than a traditional cellular network, the PCN network is particularly sensitive to transmission costs. Even if the costs of cabling to PCN cell sites were not prohibitive, availability of duct space is generally not enjoyed by new PCN operators. Key, therefore, to the implementation of PCN is access by the operator to high-frequency microwave and millimeter-wave radio capacity for linking cell sites back to switching centers.

The macrocell and microcell dimensions are such that, rather than laying costly underground cable links, millimeter-wave hops become an attractive proposition. In the United Kingdom, the DTI is allowing PCN operators to use the 38-GHz and potentially 55-GHz bands for this purpose. These links can be used for ranges up to a few kilometers. The general architecture of the PCN transmission network is shown in Figure 5.6. Cell sites are linked to higher-order nodal points by "daisy-chain" or star configurations. Clearly, the most cost-effective solution is to use the cell sites wherever possible as the first-stage transmission nodal point, and the selection and implementation of cell sites should be driven as much by the transmission connectivity (e.g., line of sight) needs as by the cellular radio engineer's requirements for radio coverage.

Point-to-point radio link equipment operating at 38 GHz is now available at low cost, and this provides the most cost-effective solution to the first stage "hop" from a cell site. Such equipment can operate on links of a few kilometers, with small (30 to 60 cm) dish radii—again, compatible with the environmental considerations of cell-site implementation. Because of the short-range nature of the links and the imperatives on low-cost implementation, relatively simple digital radio link equipment is used. Simple 2- and 4-level FSK/PSK modulation systems, while not the most bandwidth efficient, provide high reliability and low cost because of their simplicity. Link frequency planning is a challenge because of the regular grid nature of the cellular network, although in

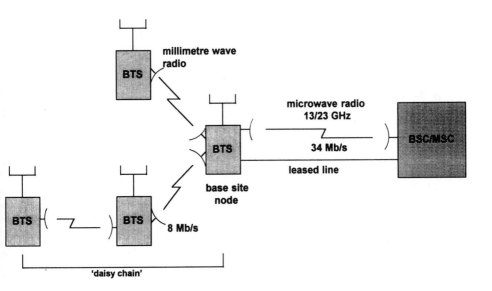

Figure 5.6 Transmission network architecture.

general the natural distortions of the grid due to practical site location issues diversifies beam patterns sufficiently to enable effective frequency reuse to be achieved.

With the DCS1800 speech coder, each transceiver requires only three 64-bit/s time slots, including signaling needs. Thus, transmission links from a cell site supporting up to 9 transceivers (around 66 full-rate speech/data channels) at 2 Mbit/s or 36 transceivers at 8 Mbit/s can be implemented. Using 38-GHz radios, a channel of 7 or 14 MHz is required and experience to date has shown an overall bandwidth requirement of around 2×200 MHz for a large city and its surrounding commuter environs is adequate for links to connect the radio cells.

In a tree-and-branch architecture, there is no route diversity from any network node, although resiliency can be improved by 1+1 hot-standby radio equipment. However, due to radio outage being primarily due to occasional adverse propagation conditions, an alternative to radio is the best means of improving network resiliency. By adding route diversity fiber (typically leased from a public telecommunications operator, or PTO), extremely reliable paths can be created due to the lack of correlation between route failure mechanisms. However, it is difficult to ensure that fibers are available at each critical node. Transmission network planning is a difficult compromise between the requirements of low cost and yet high resiliency.

5.11 DCS1800 ENHANCEMENTS

A common evolution of the GSM and DCS1800 standards was agreed as being desirable, and a joint program of work has been carried out within the ETSI SMG Technical

Committee for the evolution and enhancement of the standard. Priority tasks included completion of work on desirable features and services that were not completed as part of the Phase 1 activities, particularly for support of a Group 3 facsimile service. New supplementary services, including call waiting, call hold, multiparty, line identification, and closed user group, are being developed that will improve the PCN service offering. Optimization of the standards is taking place as results from early GSM operational systems are becoming available. The trend toward smaller cells to increase capacity, particularly in dense urban areas, requires new techniques to support microcell environments, as described in Section 5.10. In addition enhancements have been proposed to the national roaming mechanism to make its implementation more cost effective.

The PCN operators have also supported the development of a half-rate speech coder, which is a key component for still higher capacity networks. Two candidates have emerged with quality and delay similar to that of the full-rate 13-kb/s RELP GSM/DCS1800 codec, but with implementation complexities around three times as great. It is predicted that this additional complexity will be matched by the anticipated advances in semiconductor technology in the next few years.

Third-generation systems should continue the move toward systems that fully meet the requirements of personal communications by offering services and features that are demanded by the market and that can be offered economically and competitively. Developers of standards for third-generation systems have opportunities to provide technical solutions to respond to the emerging and projected market needs but will also need to take into account the significant investments that many will have made in digital infrastructure for personal, mobile, and fixed network services. Consideration should be given to exploiting the reuse of existing standardization work and systems infrastructure where this is appropriate, particularly bearing in mind the expected large populations of personal communications terminals that will be in use at the time of the introduction of the next generation of systems. ETSI has acknowledged the need for a third-generation standard and is setting requirements as the initial step toward the standardization of a universal mobile telecommunications service (UMTS) (see Chapter 12).

5.12 THE PCN VISION

It is generally accepted that the PCN vision is the provision of affordable communications with total freedom and mobility, ubiquitously available, and provided in a manner that puts the users in control of their communications. DCS1800 provides a standard to deliver the PCN vision and has enabled Mercury One-2-One to launch the world's first PCN in London during 1993 with plans to continue its implementation over the whole of the United Kingdom by around the turn of the century.

ACKNOWLEDGMENTS

The authors would like to thank their colleagues at Mercury One-2-One for their valuable contributions toward the successful implementation of PCN and help in preparing this chapter. The authors also thank the parent companies of Mercury One-2-One, Cable & Wireless, and US-West for permission to publish this work.

REFERENCES

[1] European Telecommunications Standards Institution, "Recommendations for GSM 900/DCS1800" Published by ETSI, 06921 Sophia Antipolis, Cedex, France.
[2] Ramsdale, P. A., and W. B. Harrold, "Techniques for Cellular Networks Incorporating Microcells" *IEEE Conf. PIMR 92 Boston*, Oct. 1992.

SELECTED BIBLIOGRAPHY

Balston, D. M., and R. C. V. Macario, *Cellular Radio Systems*, Norwood, MA: Artech House, 1993.

Chapter 6

Personal Communication Services in the U.K. Cellular Environment

Trevor Gill

The present development of a wide range of public mobile telecommunications services in the United Kingdom is firmly based on the success of the two competing cellular telephone networks, which commenced operation in January 1985. Since that time both Cellnet and Vodafone have experienced phenomenal growth; at the beginning of 1994 they had over 1.9 million subscribers between them and were two of the largest mobile phone operators in the world. The competitive market has led to some of the lowest prices in Europe, with tariffs held constant outside London since service began.

The existing cellular networks are based on analog mobile radio technology developed by Bell Laboratories in the United States during the 1970s. This technology was adopted, with minor modifications to suit European frequency allocations around 900 MHz, as the U.K. total access communication system (TACS). As originally designed, the system was intended to support relatively high-powered mobile equipment installed in vehicles. From the earliest days in the United Kingdom it was anticipated that portable equipment would be increasingly used on the network. As the networks have expanded, handportable equipment has become much more compact, to the point that truly pocketable phones have become available at decreasing prices. The demand for portable equipment is such that 40% of phones in use are now portables, and the number of portables registered on the Vodafone network is actually increasing faster than the total number of subscribers.

As the networks have expanded, cell sizes have decreased such that portable equipment is adequately served over a large proportion of the country. In early 1994, the Vodafone network had about 800 (analog) cell sites with more than 30,000 voice channels in service.

Although the cost of using a mobile phone continues to fall in real terms, it is still considerably greater than that of using a fixed telephone. The existing cellular networks serve a subscriber base consisting largely of business users, for whom the additional cost is justified by the benefit of mobile communication.

The challenge for the future is to bring mobile communications to the mass market. The first step in this direction is the adoption of a common standard across Europe for a second-generation digital cellular telephone system. This system is known as GSM, an acronym that began as the name of a European committee and has since been somewhat grandly renamed the global system for mobile communication. This system will gradually replace a wide range of incompatible analog cellular systems across Europe during the 1990s. The European Community set an ambitious target of July 1991 for the opening of service of GSM networks. At the end of 1993, there were 23 working GSM networks in 14 countries serving 1.5 million subscribers. This is a remarkable achievement considering the development work that still remained to be done once the specifications for the system became stable in the late 1980s.

The GSM systems will operate in the same part of the spectrum around 900 MHz currently used by the TACS system. The initial allocation in the United Kingdom is of 2×10 MHz, expanding to 2×25 MHz when the analog networks are finally phased out in the next century. Already, some countries have identified a potentially greater demand for spectrum, and another 2×13 MHz is being suggested as an expansion band for GSM in the future.

Before the GSM systems had even begun operation, the U.K. government had licensed three operators to provide personal communications services, using a spectrum allocation of 2×75 MHz at around 1800 MHz. After some discussion about the technology to be used, the operators settled on a system firmly based on GSM technology, now standardized in Europe and known as DCS1800.

So what is the difference between a PCN and a GSM cellular network? In the remainder of this chapter, we will consider the similarities and the differences and the way that personal communications services will be provided by the existing cellular operators.

6.1 GSM and DCS1800: SIMILARITIES AND DIFFERENCES

The differences in technical specifications between the GSM 900 and the DCS1800 systems are relatively small. Both are based on identical modulation and signal processing functions and can potentially offer the same range of services. The most significant differences are summarized in Table 6.1.

It can be seen that the greatest difference is in the increased spectrum available to the personal communications network (PCN) operator. This gives the operator greater flexibility in frequency planning compared to a GSM network and potentially a capability to serve a larger customer base in the long term. The GSM specification supports a range of equipment from low-power handheld to high-power mobile. The DCS1800 system has been specified from the start to support only low-power handportable equipment. This

Table 6.1
Major Differences in GSM 900 and DCS1800 Specifications

	GSM 900	DCS1800
Mobile TX band	890–915 MHz	1710–1785 MHz
Mobile RX band	935–960 MHz	1805–1880 MHz
Mobile TX peak power (maximum)		
Class 1	20W (+43 dBm)	1W (+30 dBm)
Class 2	8W (+39 dBm)	0.25W (+24 dBm)
Class 3	5W (+37 dBm)	
Class 4	2W (+33 dBm)	
Class 5	0.8W (+29 dBm)	
Mobile TX peak power (minimum)		
Class 1	+13 dBm	+10 dBm
Class 2	+13 dBm	+4 dBm
Class 3	+13 dBm	
Class 4	+13 dBm	
Class 5	+13 dBm	
RX sensitivity (portable phone)	−102 dBm	−100 dBm

removes a degree of flexibility in the range of services that can be provided compared to a GSM network. For example, a GSM network could be designed to allow handheld coverage in urban areas but full countrywide service in rural areas via the use of a power-boosting car adaptor. The slightly greater power control range of the DCS1800 equipment and other detailed differences in transmitter and receiver performance specifications too numerous to mention here give the DCS1800 system a small performance advantage when operating with a high density of subscribers.

Another small advantage of the DCS1800 system is due to the fact that multipath fading occurs more rapidly at the higher frequency for a given speed of the moving subscriber. Because the error correction system inherent in GSM and DCS1800 is better at correcting randomly spread errors than bursts of errors, it is enhanced by an interleaving system that spreads out bursts of errors. This is made more effective at the slow speeds of a pedestrian by the higher fading rate at 1800 MHz.

In contrast, the GSM operator has a considerable advantage in path loss capability, and thus in the maximum cell size that can be achieved. The free-space path loss between isotropic antennas increases by 6 dB for a doubling in frequency. Additional losses due to diffraction also increase with frequency. The empirical propagation model derived by Hata [1] suggests an additional path loss of 7.7 dB at 1800 MHz. Since the release of spectrum for mobile applications is around 1800 MHz, many organizations have carried out comparative trials at 900 and 1800 MHz. Simultaneous trials in Aalborg, Denmark [2], showed a mean additional loss of 9.8 dB. Similar trials in Mannheim and Darmstadt, Germany, yielded a mean difference of 11 dB, although in one industrial area with large buildings and little vegetation the difference was only 6.4 dB. The difference in path loss in open rural areas will be lower; Hata [1] suggests 4.2 dB, but the penetration loss of trees in wooded areas will again increase the loss at 1800 MHz. So far, we have considered only outdoor coverage, but an important aspect of any truly portable service is coverage inside buildings. Data on building penetration loss at different frequencies have been gathered by many organizations. The vast variety in building construction methods, shapes, and sizes leads to a large body of data from which it can be difficult to draw clear conclusions. On balance, it seems that there is little to choose between the two frequency bands.

Overall, it is reasonable to conclude that a network operating at 1800 MHz will have a disadvantage of around 10 dB in path loss. Although some of this disadvantage can, in theory, be regained by the use of higher-gain antennas for a given physical size, this advantage will be difficult to achieve in practice. In a portable equipment it is difficult to achieve any significant gain, because the orientation of the antenna cannot be controlled. The large collinear arrays already used for base station antennas at 900 MHz typically consist of up to 8 dipoles. Any larger array would have an excessively narrow vertical radiation pattern and would have to be carefully manufactured to have more than a 1- to 2-dB advantage. The reduced receiver sensitivity of the DCS1800 equipment is no disadvantage, because the base station power can be increased to compensate, the uplink power budget from the low-power handheld being the limiting factor. On the uplink, a class-1 DCS1800 phone has a 1-dB advantage over a class-5 GSM phone.

In conclusion, since a path loss reduction of 10dB corresponds approximately to a doubling of range, the reduced path loss at the lower frequency will allow a cellular network operating at 900 MHz to provide coverage to portable equipment with considerably fewer base stations than would be required at 1800 MHz. This is of considerable advantage to an operator attempting to roll out a network at minimum cost. It can be seen that although there are important differences in spectrum available and in the maximum cell size achievable, the two systems are technically very similar. What then is the difference between a ''personal communications'' service (PCS) and a ''cellular'' service, when even the existing analog networks provide substantial support for pocket portable phone users? The biggest differences between PCN services and cellular services and between different PCN services will be in the markets the operators choose to address and the way they target coverage and tariffs. The same technology can be used to provide a nationwide mobile service, a localized portable service, a system whose main aim is to bypass the local loop in supplying telephone service to domestic subscribers at home, or some combination of all three.

6.2 PERSONAL COMMUNICATIONS SERVICES

Stimulated by competition from the PCN operators, it is inevitable that both Cellnet and Vodafone will want to offer a range of GSM-based services targeted at different groups of customers, from the business user to the domestic telephone subscriber. The PCN operators will be aiming at the potentially large but highly cost-sensitive domestic market, by offering a lower-cost service than the traditional cellular systems. The key to success for the cellular network operator will be the ability to offer a variety of services with different characteristics and different tariffs. Some of these services will compete head-on with the PCN operators, while others will offer additional facilities for premium prices. The different services may be characterized by different coverage areas, regional and local tariffs, or even different grade of service, measured in terms of the probability of being able to make a call.

There are three fundamental requirements in the system infrastructure to support flexibility in services and tariffs. The first requirement is a means for on-line checking of subscription information, so that at the time of access the network can determine the level of service to which a particular subscriber is entitled.

The second is an on-line method of informing the user of the cost of a call before it is made. It is not unreasonable that most consumers expect to know the cost of something before they commit to buying it. In a fixed network, the operator can publish a fixed tariff, and the user is always able to determine the cost of a call in advance. If a user is in any doubt, for example, of which calls are carried at the local rate, he or she merely has to look up a list of exchanges in a telephone directory. In contrast, in a mobile network, it may be desired to set a tariff in which the calls that are charged at a local rate depend on the cell that is currently serving the subscriber. It is not practicable to publish a definitive map showing the area served by each cell, and even if it were, the user would not want to keep consulting it. If subscribers are not to become frustrated by receiving bills for expensive calls that they reasonably expected to have been cheaper, they must be able to determine the tariff for each call in advance. Once this facility is available, it will offer the network operator an enormous flexibility in setting tariffs. The operator might, for example, dynamically increase the price when a cell is heavily loaded. This represents a highly responsive method of market pricing; if the user dislikes the offered price, he or she can try again later when the congestion has been relieved and the price has dropped.

The third requirement is a suitable off-line billing system that can generate the appropriate entries in the customer's bill from a limited quantity of data logged when the call was made.

To consider the way in which some of these requirements will be achieved, we will look at a practical example. At the Communications 91 exhibition in Geneva, Vodafone announced plans for a PCS service as a microcellular network (MCN); the service has since been launched under the name MetroDigital. The following section describes the features of that network, and the way it is integrated into the full Vodafone GSM cellular system.

6.3 THE VODAFONE MCN SERVICE

The MCN service to be offered by Vodafone will be designed to offer an economical service to the user of a lightweight, class-5 GSM portable phone. The service will be available in any significant built-up area, identified as the areas shaded in yellow on an ordinance survey route planning map. At the opening of service in 1993, the southeast of England was fully covered, with rollout to most of the country over three years.

MCN tariffs will be at three levels. The "home local" tariff will apply to all calls made from a small area surrounding an address that the user nominates as his or her home location. This will usually be a home address but may be a business address. The "roamed local" tariff will apply when the user is away from his or her home location but makes a call to a nearby fixed subscriber. For example, a user whose home location is in London but who is currently in Portsmouth would pay the roamed local rate for a call to Southampton. Any other calls will be at the "national" rate, which will still be lower than the rate for all calls on the full GSM network.

The full GSM network will provide almost nationwide coverage, as the TACS networks do today, although not all areas will be covered adequately for portable phone users. The full GSM network will be available to an MCN subscriber who selects the GSM network manually or clips the phone into a car adaptor kit incorporating a power booster. Calls made in this way will be at the "GSM upgrade" tariff, which will be more expensive than what a subscriber to the full GSM service would pay for the same call.

The GSM and MCN services are, therefore, complementary. Users who make calls mainly within towns or villages with MCN coverage and only occasionally use their phones while traveling between populated areas would select the MCN service, but they do not sacrifice the availability of nationwide coverage when they really need it. Users who make frequent calls while on the road would find it cheaper to subscribe to the GSM service, with the full network available to them at all times. Users of either network will, of course, have full access when traveling abroad to the range of GSM services that will cover the majority of Europe during the 1990s. An MCN or GSM subscriber can, therefore, be reached on a single number whenever he or she is within range of a GSM network anywhere in Europe. As with any GSM subscriber, incoming calls to a user outside the home country will be charged to the caller at the normal rate for a call to the home mobile network. The additional cost due to the subscriber being abroad will be charged to the subscriber receiving the call.

6.4 GSM AND MCN INFRASTRUCTURE

The two networks will be supported by three different types of cell sites, which will be known as:

- GSM macrocells;
- Shared macrocells;
- Shared microcells.

GSM macrocells will be large cells, generally covering rural areas, and available only to subscribers to GSM service or MCN subscribers who have temporarily elected to use the

GSM network while paying a premium rate. Shared macrocells are cells that form part of the main GSM network but that provide adequate coverage to handportable equipment over part or all of a built-up area. MCN subscribers will treat these cells as part of the MCN network. Macrocells may be either omnidirectional, in which one site feeds one cell, or sectored, in which three different cells are fed by directional antennas from a single site. The vast majority of macrocells will be located at sites already used for the analog TACS network, so rollout can be rapid. Shared microcells are cells constructed specially to support handportable coverage of built-up areas not already covered adequately by a macrocell. They will generally be omnidirectional, serving an area within a radius of about 2 km in suburban districts.

To provide portable coverage of most built-up areas of the United Kingdom, it is anticipated that approximately 2,500 shared microcells will augment the approximately 1,600 macrocells fed from 750 sites.

To allow rapid installation of microcell equipment, a standard cell configuration has been designed, which uses a sufficiently small mast to minimize environmental impact and avoid the need for planning permission. Figure 6.1 shows a typical microcell site. The mast is about 15m high and will support two omnidirectional antennas. One antenna will be used for transmission, and both will be used for diversity reception. Diversity reception helps to mitigate the effects of multipath fading and is particularly important in a system designed for portable use. Without diversity, a static or slowly moving user might pause at a point where the signal on the uplink (mobile to base station) faded into a deep null. Because the fading on the uplink and downlink frequencies is not correlated, the user may be hearing a perfectly good signal on the downlink, but the call may be dropped. The use of diversity reception greatly reduces the probability of this happening.

The mast will also be capable of supporting one or more microwave dishes. Wherever possible, microcells will be connected to a nearby macrocell site by microwave link, to minimize link costs. In selection of potential microcell sites, the availability of a line-of-sight path to a nearby macrocell will be an important consideration. The base station transceivers, baseband equipment, microwave link, operations and maintenance systems, control equipment, and a battery backup power supply will all be contained in a single environmental cabin that will be sited near the bottom of the mast. All the equipment will be factory commissioned, so that on-site installation work involves little more than mounting the cabin on its concrete base and connecting up power and antennas. In urban areas, where a substantial proportion of buildings exceed a height of 15m, a similar package will be rooftop mounted on a suitable building.

6.5 SERVICE IMPLEMENTATION

Both GSM and PCN networks are based on a common set of standards controlled by the European Telecommunications Standards Institute (ETSI). The ETSI standards specify a minimum set of requirements to ensure that mobiles and infrastructure from different

Figure 6.1 Typical microcell site.

manufacturers can be configured to work together. This alone gives network operators a greater degree of choice in purchasing equipment than with earlier analog systems. In the analog systems the radio interface was standardized to allow competition in mobile manufacture, but the interfaces between base stations and switching equipment were generally proprietary. This meant that once a network operator had chosen a supplier, the operator was committed to that supplier for the life of the network. With DCS1800 and GSM, it is possible for an operator to mix switches from one supplier with base stations from another.

Although both DCS1800 and GSM networks are based on the same set of European standards, this does not imply that all networks will be identical. Within the standards there is scope for each manufacturer to develop proprietary strategies for functions such as handover. We will now consider some of the specialized functions that are required to implement an integrated set of mobile and personal communications services based on the GSM standards.

6.5.1 Advice of Tariff

We have seen that advice of the cost of the call is an important feature in any system supporting multiple tariffs. There is more than one way in which this might be provided. The simplest would be by the use of a very short recorded or synthesized voice announcement to the subscriber before a call is connected. An alternative is by the use of the GSM cell broadcast short-message service (CBSMS). This service allows a text message of up to 93 characters to be broadcast to every mobile in a cell. This may be able to be used to deliver some indication of charge rates to the user, who would be able to display the information on the handset. While the CBSMS is a broadcast to all mobiles, GSM also provides for the short-message service (SMS), which allows text messages of up to 160 characters to be addressed to a specific mobile. Unlike CBSMS, SMS uses an acknowledged protocol, so that successful delivery of the message to the mobile is guaranteed. Since a message can be addressed to a specific mobile, use of SMS would allow more complex tariffs to be applied than use of the broadcast version.

6.5.2 Trunk Reservation

Where a range of different services is provided over the same network infrastructure, it is important to ensure that the resources are shared equitably among users of the different services. All telephone systems, whether mobile or fixed, rely heavily on the idea of trunking. If enough telephone lines were provided between London and Manchester for all subscribers in London to make phone calls to Manchester simultaneously, a very expensive resource would be lying idle for most of the time, and telephone calls would be astronomically priced. Instead, the network designer relies on the fact that only a small proportion of Londoners are phoning Manchester at any time. Instead of designing this system for a worst case, which will never occur, the system is designed on the basis of probability. Teletraffic theory enables the designer to calculate how many lines are required, assuming how many Londoners on *average* will be phoning Manchester, and what probability of being able to make a call will be acceptable to the customer. Such a service would generally be designed so that a call attempt would be successful something like 98% of the time during the busiest hour of the day. Only on one attempt in 50 would the user receive a recorded message saying, "Lines to Manchester are busy, please try later." The probability of a call failing is known as the *grade of service*, so this system would be described as being designed for a grade of service of 2%. The ratio of the average number of phone calls attempted to the number of lines that must be provided is a measure of the trunking efficiency. Trunking efficiency increases rapidly with the number of lines, especially when the number of lines is small, so that two lines can support far more than twice the number of subscribers supported by one line.

In the initial stages of rollout of a PCS, it is likely that each cell will incorporate only one transceiver, supporting a maximum of seven simultaneous phone calls. As traffic on the network increases, additional equipment will be added to each cell. At 2% grade of service and assuming Erlang B statistics, seven lines can support an average of 2.9 simultaneous phone calls. If this small number of lines were to be split to serve two different services, the total capacity would be greatly reduced. Four lines can support 1.1 calls, three lines only 0.6, so the two services together would support only 1.7 simultaneous calls on average. It can be seen that splitting the lines available between the two services would make inefficient use of the available resources. On the other hand, if the full resource were available to users of either the GSM or MCN service, and the system should become overloaded, there is the possibility that a GSM user who has paid a higher subscription in order to achieve a more comprehensive service might be denied access due to a large volume of traffic generated by MCN users paying lower tariffs.

To solve this problem, Vodafone has patented a system known as Trunk Reservation. This system is designed to control the grade of service offered to users of a number of different services that share the same infrastructure. Let us consider a cell designed to provide n different services, (e.g., service 1 = GSM, service 2 = MCN). If there is a request for the use of a voice channel and a channel in the cell is free, the system will decide whether to grant the request depending on the service requested, the number of channels free, and the value of a random number, R. The random number may be uniformly distributed between 0 and 1. A new random number is generated for each access request. Table 6.2 contains values for X_{ij}.

Access to the network is granted only if the random value R is greater than or equal to the value X_{ij}, where i corresponds to the number of free channels and j to the service requested. If more than some number of channels Q are free, all access attempts will be granted. With this system, suitable choice of the coefficients X_{ij} enables the operator to adjust the grade of service offered to each service as appropriate.

6.5.3 Service Separation

For two or more separate services to be supported on the same network, a means must be provided to allow the network to determine which users may have access to any given

Table 6.2

Trunk Reservation Coefficients

Free Channels	Service Type		
	1	*2*	*3 . . .*
1	X_{11}	X_{12}	
2	X_{21}	X_{22}	
3	X_{31}	X_{32}	
$\vdots$	$\vdots$	$\vdots$	
$\geq Q$	1	1	

cell. In the Vodafone network, the system will initially identify a user as requesting MCN or GSM service by that user's classmark. A class-5 phone will receive MCN service, while phones in classes 1 through 4 will receive GSM service. In a GSM network, subscription details are recorded not in the phone itself but in a removable *subscriber identity module* (SIM). This will usually be implemented as a credit card-sized "smart card." The results of using SIMs containing GSM and MCN subscription information in different mobiles are shown in Table 6.3. The combination of an MCN subscription and a phone in class 1, 2, 3, or 4 may, of course, be obtained either by the MCN subscriber using his or her smart card in a different phone or by inserting a class-5 phone into a power-boosting car adaptor.

Mobility management in a GSM network, that is, the ability of a network to know where to find a given mobile, is handled by a procedure known as *location updating*. The network is broken down into a number of location areas. Each base station radiates a signal containing the location area identity (LAI) of the area to which it belongs. Whenever a mobile camps on a cell with a different LAI, it carries out the location update procedure to inform the network that it has moved into a new area. The size of a location area is determined by a trade-off between the volume of signaling traffic generated by paging requests and the volume of location update traffic. The two extremes of this trade-off are easily understood. If the entire network were treated as a single location area, the network would have no idea where any mobile was located. In the event of an incoming call to the network, a paging message would have to be transmitted in every cell. This would

Table 6.3

Service Available for Each Subscription and Classmark Combination

SIM Subscription Type	Mobile Class	
	1–4	*5*
GSM	Full GSM service at GSM tariff	Service available only from MCN cells, GSM tariff
MCN	Full GSM service at premium tariff	Service available only from MCN cells, MCN tariff

lead to an excessive loading of the control channel with paging traffic. At the other extreme, every cell could be treated as a separate location area. In that case, every time a mobile crossed from one cell to the next, it would generate a location update. Paging traffic would be minimized, because the network would only have to page a mobile in one cell, but the control channel would be heavily loaded with location update traffic. The practical solution, of course, lies somewhere in between, with a group of cells in each location area.

The location-updating mechanism can be used to support separation of cells providing MCN service from those that provide only GSM service. GSM-only cells will be given a different location area from shared cells in the same locality. A class-5 mobile attempting a location update on a GSM-only cell will have its location update rejected. This allows the system to restrict MCN users to a subset of cells in the network. The allocation of location areas is illustrated in Figure 6.2.

Use of techniques such as those described here offers the cellular network operator the necessary flexibility to be able to offer a range of personal communications services to suit the needs of its customers and to respond to competition in the fast-changing marketplace for mobile communications in the 1990s.

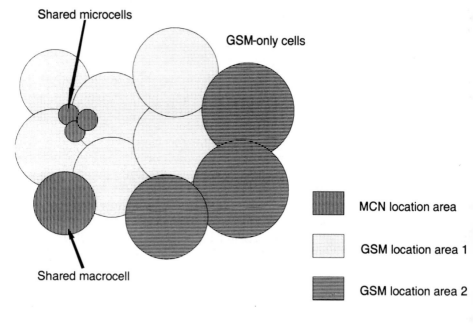

Figure 6.2 Allocation of location areas.

REFERENCES

[1] Hata, M., "Empirical Formula for Propagation Loss in Land Mobile Radio Services," *IEEE Trans. on Vehicular Technology*, Vol. VT-29, No. 3, Aug. 1980, pp. 317–323.

[2] Morgensen, P.E., P. Eggars, and C. Jensen, "Urban Area Radio Propagation Measurements for GSM/DCS1800 Macro and Micro Cells," *Proc. 7th Int. Conf. Antennas and Propagation*, April 15–18, 1991, IEE Conference Publication No. 333, IEE, London, 1991, pp. 500–503.

Chapter 7
Wireless Local Area Networks

Kaveh Pahlavan

This chapter provides an overview of wireless data networks from the dual perspectives of market trends and applications and describes various technologies and architectures used in the implementation of wireless local-area networks (WLLANs).

While it is true that the major thrust of telecommunications is toward multimedia services, the existing infrastructure of communications networks is still fragmented. Today we have wired private branch exchanges (PBXs) for local voice communications within office complexes, the public switched telephone network (PSTN) for wide area voice communications, wired local-area networks (LANs) for high-speed local data communications, packet-switched networks and voice-band modems for low-speed wide-area data communications, and a separate cable network for wide-area video distribution. The process of setting standards for various areas of communications has been similarly fragmented. The standards in voice transmission technology have evolved within the operating companies, while the standards for voice-band data modems have been developed by the CCITT, and the standards for local-area networks by IEEE-802 and ISO. This separation has come about because each individual network was designed to meet the requirements of a particular type of service, be it voice, data, or imagery and video.

The same pattern of separation exists in the wireless information network industry as well. The new-generation wireless information networks are evolving around either voice-driven applications, such as digital cellular, cordless telephone, and wireless PBX,or data-driven networks, such as WLLANs and mobile data networks. While it is true that all the major standards initiatives are focused on integration of services, one still sees a separation of the industrial communities that participate in the various standards bodies. That is, we see GSM, IS-54, TR45.3, DECT, and others supported primarily by representatives of the voice-communications industry, while IEEE 802.11, WINForum, and HIPER-LAN are supported primarily by those with interest in data communications. Figure 7.1

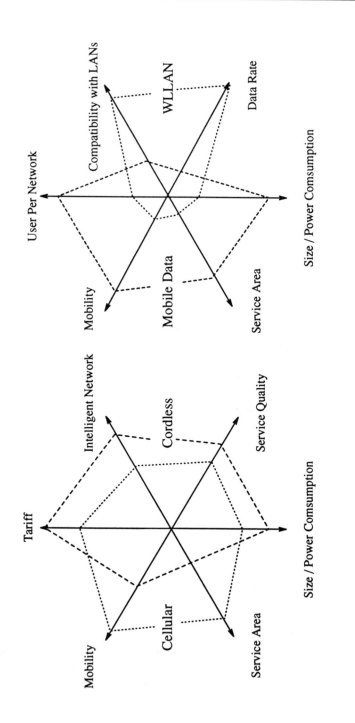

Figure 7.1 Various aspects of the voice and data services offered by wireless information network industry. Mobile cellular is compared with the cordless personal service and WLLAN with mobile data.

shows various dimensions of the voice and data communication industries and compares the local cordless personal communication with the wide-area cellular and the WLLANs with the wide-area low-speed data services.

Although future personal devices may be designed as integrated units for personal computing as well as for personal voice and data communications, the wireless access supporting different applications may use different frequency bands or even different technologies. A personal communications service (PCS) may use a shared wideband-code-division multiple access (CDMA) band, and digital cellular service another time-division multiple access (TDMA) or CDMA band; while low-speed mobile data are carried in the gaps between bursts of voice activity, and high-speed local-area data uses another shared wideband channel. At the same time, the various services may all be integrated in a metropolitan-area or wide-area network structured with asynchronous transfer mode (ATM) switches. The future direction of this industry depends on technological developments and a maturity in the spectrum-administration organizations, which must understand the growing massive demand for bandwidth and must in turn develop strategies allowing a fair sharing of increasingly scarce bandwidth. Just as governmental agencies restrict abusive consumption of other limited natural resources, such as water, appropriate agencies will have to protect the spectral resources needed for wireless information networks. After all, the electromagnetic spectrum is a modern natural resource supporting the ever-widening array of telecommunications services, which are becoming an increasingly important part of the fabric of our personal and professional lives.

7.1 WIRELESS DATA: MARKET AND USER PERSPECTIVES

From the data user's perspective, the minimum satisfactory requirement for personal communications is low-speed access in wide areas and high-speed access in local areas. The low-speed wide-area access will serve a variety of short-message applications, such as notification of electronic or voice mail. The local-area access will support high-speed local applications, such as long file transfers or printing tasks. In the current literature, low-speed wide-area wireless data communication is referred to as mobile data, and local high-speed data communication systems are called wireless LANs, or WLLANs. The relationship between WLLAN and mobile data services is analogous to the relationship between PCS and digital cellular services. While PCS is intended to provide high-quality local voice communication, the digital cellular services are aimed at wider-area coverage with less emphasis on the quality of the service.

7.1.1 Mobile Data

Mobile data radio systems have grown out of the success of the paging-service industry and increasing customer demand for more advanced services. Today 100,000 customers in the United States are using mobile data services, and the industry expects 13 million users by the year 2000. This could account for 10% to 30% of the revenue of the cellular radio industry. Today, mobile data services provide length-limited wireless connections with in-building penetration to portable users in metropolitan areas. The future direction is toward wider coverage, higher speeds, and capability for transmitting longer data files.

The data rates of existing mobile radio systems are comparable to voice-band modem rates (less than 19.2 Kbps). However, the service has a limitation on the size of the file that can be transmitted in each communication session. The coverage of the service is similar to the mobile radio services with the difference that the mobile data service must have in-building penetration. Mobile radio users typically use the telephone unit inside the vehicle and usually while driving. Mobile data users typically use the portable unit in a building and in a fixed location. Therefore, in-building penetration is an essential feature of mobile data services.

Mobile data services are used for transaction processing and interactive, broadcast, and multicast services. Transaction processing has applications such as credit card verification, taxi calls, vehicle theft reporting, paging, and notice of voice or electronic mail. Interactive services include terminal-to-host access, remote LAN access, and electronic games. Broadcast services include general information services, weather and traffic advisory services, and advertising. Multicast services are similar to subscribed information services, law enforcement communications, and private bulletin boards.

Other low-speed data products use voice-band modems over existing radio channels originally designed for voice communications. A group of these modems use the land-mobile-radio bands around 100 to 200 MHz normally used by handheld radios for low-speed local data communications in or around buildings. Other groups are using voice-band modems over the analog cellular telephone network to provide wide-area data communications for mobile users without any restrictions on the length of the connection.

7.1.2 WLLANs

Today, most large offices are equipped with wiring for LANs, and the inclusion of LAN wiring in the planning of a new large office building is done as a standard procedure, along with planning for telephone and electric-power wiring. The WLLAN market will very likely develop on the basis of the appropriateness of the wireless solution to specific applications. The targeted markets for the WLLAN industry include applications on manufacturing floors, in offices with wiring difficulties, in branch offices, and in temporary offices. In manufacturing facilities, ceilings typically are not designed to provide a space for distribution of wiring. Also, manufacturing floors are not usually configured with walls through which wiring might otherwise be run from the ceiling to outlets. Underground wiring is a solution that suffers from expensive installation, relocation, and maintenance costs. As a result, the natural solution for networking on most manufacturing floors is wireless communications. Other wide indoor areas without partitioning, such as libraries and open-architecture offices, are also suitable for application of WLLANs. In addition, buildings of historical value, concrete buildings, and buildings with marble interiors all pose serious problems for wiring installation, leaving WLLANs as the logical solution. WLLANs are also the choice of unwired small offices, such as real estate agencies, where only a few terminals are needed and where there may be frequent relocations of equipment to accommodate reconfiguration or redecoration of the office space. Temporary offices, such as political campaign offices, consultants' offices, and conference registration centers, provide another set of logical applications of WLLANs. The WLLAN industry expects to capture 5% to 15% of the LAN market in the near future.

Although the market for PCS is not growing as it has in past years, the market for portable computers such as laptops, pen-pads, and notebook computers is growing rapidly. Of greater importance to the wireless data communication industry, the market for networked portables is growing much faster than the market for portable computing. Obviously, wireless is the communication method of choice for portable terminals. Mobile data communication services provide a low-speed solution for wide-area coverage. For high-speed and local communications, a portable terminal with wireless access can bring the processing and database capabilities of a large computer directly to specific locations for short periods of time, thus opening a horizon for new applications. For example, one can take a portable terminal into classrooms for instructional purposes or to hospital beds or accident sites for medical diagnosis.

7.2 FREQUENCY ADMINISTRATION ISSUES

The principal technological problems for implementation of wireless LANs are (1) data rate limitations caused by the multipath characteristics of radio propagation, (2) the difficulties associated with signal coverage within buildings, and (3) the need for low-power electronic implementations suitable for portable terminals. These technical difficulties can be resolved, and effective solutions are being developed for all of them. The greatest obstacle for the wireless multimegabit data communication services is the lack of a suitable frequency band for reliable high-speed communication. The existing industrial, scientific, and medical (ISM) bands assigned for multiple-user applications are suitable for WLLANs; however, they suffer from careless users who cause unnecessary interference to other users, and they are restricted to spread-spectrum technology. More widespread use of high-speed wireless data communication technology will depend on cooperation from frequency administration organizations in providing wider bandwidth allocations without restriction on the adopted technology and in administering rules and etiquette for ethical use of these bands.

At frequencies around several gigahertz the technology is available for implementation with a reasonable size, power consumption, and cost. Moving to higher frequencies is the solution for the future. As the frequency increases, the prospect for obtaining a wider bandwidth from spectrum regulation agencies will improve. However, with today's technology, implementation at a few tens of gigahertz with reasonable product size and power consumption is challenging, particularly when wideband portable communication is considered. At higher frequencies signal transmission through walls is more difficult. For frequencies around a few tens of gigahertz, the signal is mostly confined within the room. This feature is advantageous in certain applications, such as military applications, where confinement of the signal within a room or building is a desirable security feature. Also, at higher frequencies the relationship between cell boundaries and the physical layout of the building is more easily determined, facilitating the planning of cell assignments within the building. The technology at higher frequencies is highly specialized and not commonly available within the computer industry. This has encouraged joint ventures among semiconductor, radio, and computer companies to develop new products at these frequencies.

7.3 EXISTING WLLANS

The existing technologies for WLLANs are the licensed cellular systems operating at 18 to 19 GHz, unlicensed spread-spectrum systems operating in ISM bands, and diffused and directed-beam infrared (IR) systems. Table 7.1 summarizes the features of the systems currently available in the market. Diffused IR systems provide moderate data rate and coverage. It is suitable for moderate-sized offices, short-distance battery-oriented applications (portable to printer), and environments with significant amounts of radio interference. Directed-beam IR offers higher speed with reasonable coverage for applications with fixed terminals. It is suitable for large-file transfers between mainframes and servers as well as large open offices with many fixed terminals. WLLANs at tens of gigahertz are suitable for high-speed communications in large partitioned areas, such as large offices and some libraries. The spread-spectrum systems provide the largest coverage and are suitable for penetration through building floors. Spread-spectrum technology is suitable for small-business applications where a few terminals are distributed over several floors of a building. An important factor in selection is the ease of installation and flexibility.

Table 7.1

Comparison of Radio and IR Systems

Technique	Diffused Infrared	Directed Beam Infrared	Radio Frequency	Radio-Frequency Spread Spectrum
Data rate	1 Mbit/s	10 Mbit/s	15 Mbit/s	2–6 Mbit/s
Mobility	Good	None	Better	Best
Detectability	None	None	Some	Little
Range	70–200 ft	80 ft	40–130 ft	100–800 ft
Frequency/ wavelength	$\lambda = 800\text{--}900$ nm	$\lambda = 800\text{--}900$ nm	$f = 18$ GHz	$f = 0.9, 2.4, 5.7$ GHz
Radiated power			25 mW	< 1 W

of the software to accommodate various protocols for interconnecting with the backbone wired LANs. The evolving next generation of WLLANs is designed to be incorporated into the laptop, notebook, and pen-pad computers, requiring significant reduction of the size and power consumption for existing LAN products. It will be ideal if these devices also provide low-speed wireless access for wide-area coverage.

7.4 RADIO PROPAGATION IN THE OFFICE ENVIRONMENT

The dependence of a wireless system on effective radio propagation leads to a requirement for measurement and modeling of the channel characteristics. The results of measurements and modeling are used for assessing feasibility of a communication technique at a given operating frequency.

Radio propagation in an office environment is complicated. The shortest direct path between transmitter and receiver is usually blocked by walls, ceilings, or other objects; therefore, the radio waves travel from transmitter to receiver via many paths with various signal powers. The transit time of the signal along any of the various paths is proportional to the length of the path, which is in turn determined by the size and the architecture of the office and the location of the objects around transmitter and receiver. The strength of each such path depends on the attenuation caused by passage or reflection of the signal from various objects along the path. Figure 7.2 shows a sample of the measured channel impulse response and the associated frequency response in a typical indoor radio environment. The impulse response shows the arrival of multiple paths. The frequency response shows the received power drops of the order of 30 to 40dB at selected frequencies. This extensive power drop is referred to as fading. If the location of the transmitter or the receiver is changed, or some object moves close to the transmitter or the receiver, the multipath condition and the frequencies selected for fading will change.

The normalized averaged received power as a function of delay is referred to as the delay-power spectrum. The square root of the second moment of the delay-power spectrum function is the root-mean-square (RMS) multipath spread, which is used as a measure of the multipath delay spread of a channel. Several statistics of the RMS spread in the indoor radio channels are available in the literature [1–7]. In most indoor radio environments, the RMS multipath spread at maximum distances of around 50m to 100m is less than 100 ns.

Multipath causes the received power to fluctuate statistically as a terminal is moved from one location to another. The average received signal power is proportional to the inverse of the distance between transmitter and receiver, raised to a certain power. This power factor is referred to as the distance-power gradient. The distance-power gradient multiplied by 10 shows the power loss in decibels per decade increase in the distance. In free-space radio propagation, the distance-power gradient is 2, which means the received power decays with the inverse of the square of the distance between transmitter and receiver, or the power decays at the rate of 20 dB per decade of distance. The distance-

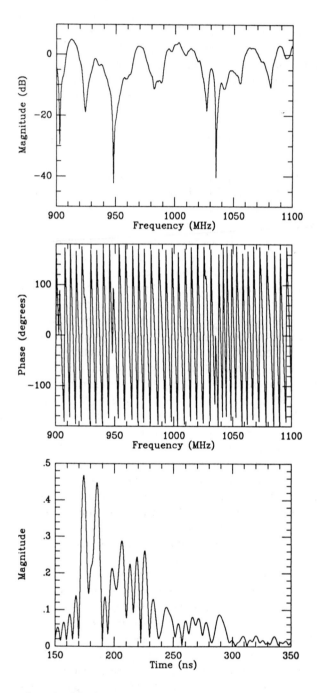

Figure 7.2 A measured sample of the time and frequency response of an indoor radio channel.

power gradient inside an office depends on the architecture of the office and the materials used in constructing the building. The available intraoffice measurements for center frequencies of around 1 GHz show power factors ranging from less than 2 up to values as large as 6 [8], corresponding to values from less than 20 dB per decade of distance up to 60 dB per decade. The smaller values correspond to hallways, which act as wavegu- ides for radio propagation, and the large values are related to buildings with metal partitioning. In the mobile radio environment, the general rule for open areas is 40 dB per decade. This value changes from one area to another if terrain characteristics are different.

The results of measurement of the statistics of the amplitude fluctuations in indoor radio channels in several rooms where the line of sight (LOS) is obstructed by walls suggest a Rayleigh distribution [9]. In a single room, a strong LOS path exists between the transmitter and the receiver, and the statistics of the received signal are modeled by either the log-normal or Rician distribution. In mobile radio channels, the local variations of the amplitude caused by shadowing are modeled as Rayleigh, and fluctuations of the amplitude in larger areas are modeled as log-normal.

The performance analysis for wideband data communication systems requires a mathematical model and a computer simulation to regenerate the profiles of the channel at different locations of a building. The computer simulation of the indoor radio propagation is done either by an approximate solution to Maxwell's equations or by the use of statistical models developed from measurements of radio propagation.

For deterministic indoor radio propagation, an approximate solution is obtained by the ray-tracing algorithm. A software package with window interface using a two-dimensional ray-tracing algorithm is described in [2]. Figure 7.3 shows a typical indoor radio channel, the trace of the rays for a specific location of the transmitter and the receiver, and the simulated channel impulse response. Further work on three-dimensional radio propagation modeling and direct solution of Maxwell's equations are currently in progress.

In wideband measurements the channel is represented by either the time response or the frequency response. In either case, the measured channel consists of samples of the channel response, which are typically on the order of several hundreds of points. To provide a reasonable coverage of a building, measurements should be taken in hundreds or thousands of locations. To observe the effects of local movements, we need several measurements at each location. As a result, the database created by measurements in a single building will require thousands of files occupying several megabytes of disk space on the computer. One way to look at the statistical modeling is that these models compress all this information into a few statistical parameters that can then be used by a programmer to regenerate a similar set of profiles on the computer. The statistics of the model are determined so that certain parameters, such as power or the RMS multipath delay spread of the profiles, fit those of the large-measurement database. Various methods of regenerat- ing the time response of the indoor radio measurements are described in [2–7]. A statistical

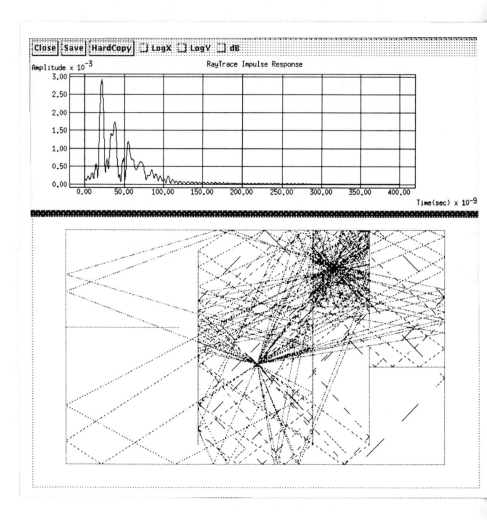

Figure 7.3 Implementation of the ray-tracing algorithm in a small office environment. All the traced paths and the overall impulse response of the channel are shown.

autoregressive model for regenerating the frequency response of indoor radio propagation is reported in [10].

7.5 DATA RATE LIMITATIONS FOR RADIO NETWORKS

LAN users strive for higher data rates; the 1-Mbps first-generation LANs are now replaced by the 10-Mbps Ethernet and 100-Mbps fiber-distributed data interface (FDDI) products are entering the market. The IEEE 802 committee is considering the next generation of FDDI at 650 Mbps. If WLLANs are to influence the LAN market, their data rate limitation has to be discovered.

The maximum symbol transmission rate for a communication system is bounded by multipath delay spread of the channel. As the symbol transmission rate increases, the duration of the transmitted symbols becomes smaller with respect to the multipath delay spread of the channel. As a result, the pulses arriving from different paths associated with a symbol interval will further interfere with the pulses arriving from different paths associated with the neighboring symbol intervals. As the data rate increases, the intersymbol interference (ISI) created by the multipath delay spread reduces the irreducible error rate of the modem. The irreducible error rate is the rate that can not be further reduced by increasing the transmission power. For any modulation technique, the irreducible error rate determines the maximum attainable data rate.

For a simple binary modulation technique, such as binary phase shift keying (BPSK) with omnidirectional antennas, the maximum data rate can be approximated by around 10% of the inverse of the maximum RMS multipath delay spread of the channel. In most indoor areas within the range of 50m to 100m, the RMS multipath spread is less than 100 ns, resulting in data rate limitations of around 1 Mbps. Results of extensive computer simulations based on measurements made in several manufacturing floors confirms the validity of this approximation. The estimated data rate of 1 Mbps is rather low for WLLANs.

These estimates improve significantly by providing diversity channels for the receiver, employing modulation and coding techniques with greater bandwidth efficiency, using multirate modems, or implementing adaptive equalization techniques.

The common methods for increasing the data rate are using antenna diversity and equalization. Diversity is usually provided with multiple transmitter and/or receiver antennas. In the indoor radio channels, diversity can also be provided by using sectored antennas [11,12]. A sectored antenna at the receiver selects the received signal arriving from specific directions. Ideally, the strongest path at the receiver is selected, and all the other paths arriving from other directions are eliminated. Elimination of unwanted paths reduces the delay spread and consequently provides an opportunity for signaling at higher rates. An adaptive equalizer is an adaptive filter at the receiver whose frequency response adapts to the inverse of the frequency response of the channel. Linear equalizers, commonly used in voice-band modems, are not effective in frequency-selective fading multipath channels, and the decision feedback equalizer (DFE) is normally used in these channels. Multipath causes ISI in the modems and so degrades performance. A DFE can isolate the arriving paths and use them as a source of internal or implicit diversity to actually improve the performance. As the data rate increases over some fractions of the inverse of the RMS multipath delay spread, the performance of the DFE modem will improve. As the RMS multipath delay spread approaches the symbol duration, the performance starts to degrade again [13,14].

Figure 7.4 shows the probability of outage versus data rate in a 30m-by-30m room for a BPSK modem with and without a DFE and compares the results with those of the modem with a sectored antenna. The outage probability represents the percentage of locations in an area for which the probability of error for the terminal exceeds a certain threshold. Radio propagation is modeled by using a two-dimensional ray-tracing algorithm. These results indicate that both sectored antenna and DFE can increase the data rate limitation of the LOS indoor radio channel by an order of magnitude. In a single-room environment, where an LOS path always exists between the transmitter and the receiver, the sectored antenna is more effective than the DFE. Figure 7.5 shows results similar to those of Figure 7.4, obtained from simulations in the multiroom indoor area shown in Figure 7.3. In a multiroom indoor area, where in most locations LOS is obstructed by walls or other objects, the DFE is more effective than the sectored antenna [12]. For the DFE, similar results are obtained from performance analysis over measured indoor radio channels [14]. A sector antenna is incorporated in the modems operating at 18 to 19 MHz to achieve an uncoded data rate of 15 Mbps with a four-symbol frequency-shift keying (FSK) modem. Using an adaptive linear equalizer is common in voice-band data communication modems, and the adaptive DFE has been implemented for troposcatter and other radio modems. However, implementation of an adaptive DFE for high-speed local packet communication systems is challenging.

The maximum data rate for a modem is selected so that for a given threshold for an acceptable performance, the modem stops proper operation only in the given percentage

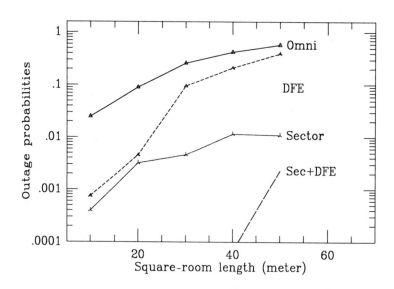

Figure 7.4 Probability of outage versus data rate for BPSK and BPSK/DFE modems with omnidirectional and sectored antennas in a 30m × 30m room.

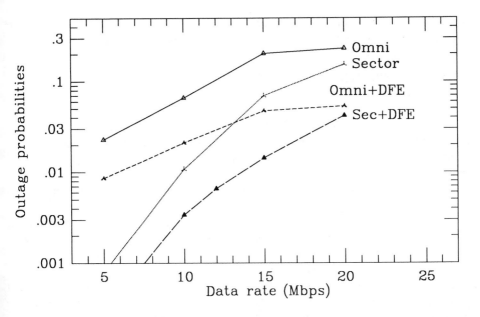

Figure 7.5 Probability of outage versus data rate for BPSK and BPSK/DFE modems with omnidirectional and sector antennas in the small indoor area shown in Figure 4.

of locations specified by the outage rate. In many of the locations where the modem performs properly, the data rate can be increased well above the operating data rate. A multirate modem assigns another higher rate of operation for those locations. The modems operate at the high data rate in most of the locations in the area where the multipath fading allows the high data rates. In locations with low signal-to-noise ratio caused by multipath fading, the data rate is decreased to adjust the error rate to an acceptable level. The work in [15] suggests that an order-of-magnitude improvement in the data rate is achievable if an optimum dual-rate modem is used. This concept is used in ISM-band WLLANs to achieve a 6-Mbps data transmission rate using spread-spectrum wireless technology. Under normal conditions, the modem operates with 16-phase shift keying (PSK) modulation with 4 bits per symbol. If needed, the number of symbols is reduced to eight and subsequently to four, which reduces the data rate to 3/4 and 1/2 respectively.

These data rates could be further increased by using more complex modulation and coding techniques to increase the number of bits per symbol transmitted. Using Trellis-Coded Modulation (TCM), today's voice-band modems have achieved 8 bits per symbol, to provide a data rate of 19.2 kbps with a channel signaling rate of 2.4 ksymbol/s. This rate is four times more than the quadrature phase-shift keying (QPSK) or Gaussian minimum shift keying (GMSK) adopted for the personal and mobile radio communication standards. However, radio modems operate at higher frequencies, and the radio channel is far more complicated than the telephone channel. As a result, implementation of these high-rate voice-band modem techniques is not straightforward with today's technology.

7.6 SPREAD SPECTRUM FOR WLLANS

Ever since the FCC announced ISM bands in May of 1985, various spread-spectrum commercial products, from low-speed fire safety devices to high-speed WLLANs, poured into the market. Today, both voice-driven and data-driven wireless information network industries are involved in spread-spectrum applications, but their points of view toward spread spectrum are quite different.

The voice-driven digital cellular and personal communications industries are considering CDMA (code division multiple access) spread spectrum as an alternative to TDMA (time division multiple access). All the studies in this industry are directed toward the bandwidth efficiency and design complexity of CDMA as compared with the TDMA. Since the interface to the telephone network at different locations is an important aspect of the wireless voice-driven services, manufacturers of digital cellular and personal communications devices must wait for the finalization of standards before they market their products. In contrast, to develop and market a WLLAN, manufacturers only need a suitable band for high-speed data communications, and the existence of a standard is not necessary. WLLANs are using spread spectrum because the first bands available for high-speed data communication were ISM bands, which are specifically assigned to spread-spectrum technology.

At the 902- to 928-MHz ISM band, 26-MHz bandwidth is available, with a minimum bandwidth expansion rate of 11. The designers of WLLAN products at this band usually sacrifice bandwidth expansion to achieve higher data rates. That way, one can achieve a data rate of around 2 Mbps with a commonly used QPSK modem. The ISM bands at higher frequencies offer wider bands and consequently higher data rates. Products operating at a data rate above 5 Mbps use the 2.4- and 5.7-GHz ISM bands. At those frequencies, the coverage of the signal is more restricted.

An important feature of the spread-spectrum technology for the WLLAN industry is the anti-multipath nature of the technique that increases the coverage and reliability of the modem. The spread-spectrum systems in the 910-MHz bands can cover several floors of a building, a feature that is not matched by other technologies. Today, an important issue for the WLLAN industry is to provide services for the portable devices that require low electronic power consumption for battery operation. In ISM bands, slow frequency hopping is considered for this purpose.

7.7 IR NETWORKS IN THE OFFICE ENVIRONMENT

Several features of IR communications are well suited for wireless office networks. Transmitters and receivers for IR systems require light-emitting diodes (LEDs) and photosensitive diodes. These diodes are inexpensive compared to RF equipment and are cheaper to install than wired systems. IR transmissions do not interfere with existing RF systems and do not come under FCC regulations. The IR signal does not penetrate walls, thus providing privacy within the office area. The only way for IR signals to be detected outside the office is through windows, which can be covered with curtains or shades. In addition to privacy, this feature of IR systems allows concurrent usage of similar systems in neighboring offices without any mutual interference. Therefore, in a cellular architecture all units can be identical, as opposed to RF connections, in which the operating frequencies of neighboring cells have to be different.

IR WLLANs are implemented using either diffused or directed beam radiation [16,17]. A diffused IR system does not require direct LOS between the transmitter and the receiver; the receiver can collect a transmitted signal by way of multiple reflection from the walls, ceiling, or other objects in the room. High-speed diffused IR can not cover an area without the help of active or passive repeaters. The directed beam focuses the radiated power toward the receiver resulting in an increase of the received power and control of the multipath. The additional received power will improve the coverage of the system. Controlling the multipath will increase the supportable data rate. The adjustments required for operation rule out this approach for some applications.

There are three main limitations that apply to IR communications. The first is due to the transient time of the IR devices: the rise time and the fall time of inexpensive LEDs limit the data rate to around 1Mbps. The other two limitations are interference from ambient light sources and the multipath characteristics of the channel for diffused communications.

The infrared content of ambient light can interfere with IR radiation and, if extensive, can overload the receiver photodiode and drive it beyond its operating point. Three sources of ambient light are daylight, incandescent illumination, and fluorescent lamps, all of which potentially interfere with IR communications. Fluorescent light is the common method of lighting in office environments and poses the most serious problem for IR communications. Fluorescent light normally has a small amount of IR radiation and during turn-on time emits a 120-Hz interfering baseband signal rich in harmonics that may reach up to 50 kHz [16]. The effects of ambient light are reduced by modulating the transmitted IR signal. The modulation carrier frequency should be at least several hundred kilohertz to avoid being compressed by fluctuation of the ambient light. For high-speed baseband communication Miller coding is usually adopted. Since a small portion of the power of a Miller-coded signal resides at low frequencies, the effects of low-frequency interference caused by ambient light is minimized. For diffused IR, the data rate limitation caused by the ambient light sources is secondary to the data rate limitations due to multipath characteristics of the channel.

Diffused IR is a multipath channel similar to a radio channel. The multipath causes a time dispersion of the transmitted symbols, and the resulting ISI restricts the maximum digital transmission rate. As in radio propagation, as room dimensions become larger, the multipath spread is increased, and the supportable bit rate is decreased. Results of computer simulations, using only the only the first reflections, suggest that the theoretical limitation for the baseband transmission rate with diffused IR is 260 Mb-meter/s [16]. Results of more elaborate simulations for reflections up to fifth order provide 60 Mb-meter/s as a realistic upper bound for data rate limitations [18]. Therefore, for a room with a length of 10m, one expects to be able to achieve a transmission rate of 6 Mbps, if multipath is the only limitation on data rate . During the past decade various modulation techniques have been examined for use on wireless optical LANs. The most popular transmission methods are pulse duration modulation, pulse position modulation, and baseband pulse transmission.

7.8 TOPOLOGIES

The common topologies used in radio networks are centralized, distributed, and multihop configurations. Figure 7.6 shows all three architectures.

7.8.1 Centralized Architectures

In a centralized architecture, shown in Figure 7.6(a), communication among all terminals is through a central control unit. With the centralized topology, coverage of the wireless network is double that of point-to-point coverage. In WLLANs, the central control unit will also serve as the obvious node to connect to the backbone wired LAN. If power control is applied to a radio network, a centralized architecture is necessary. Power control is an effective method of minimizing the radiated power of the individual terminals. Minimization of the radiated power is an effective method of conserving power in the terminals and controlling the interference. Control of the power consumption in battery operation increases battery life, which is an essential feature for portable users. The interference control in WLLANs is addressed in the next section.

Another advantage of the centralized architecture is that the control module can be located in an appropriate position to facilitate the radio wave propagation between the users and the control unit. In IR networks, the central unit is often installed on the ceiling to provide a better coverage. For frequencies on the order of 10 GHz, it is usually recommended that the central unit be located at a high location on a wall. A drawback of the centralized topology is the existence a single failure point. If the central control module fails, the entire network will fail. A fair measure for occupancy of the allocated channel is hertz per second, which measures the time duration of the usage of the bandwidth. This provides a meaningful measure to gauge the fair usage of a common bandwidth among multiple users. Another measure of the quality of data communication is the delay characteristics of the network. In a centralized architecture, each packet is store-and-forwarded twice before arrival at the destination terminal. As a result, the store-and-forward delay and the period of time that the given bandwidth is occupied (hertz per second) are both doubled.

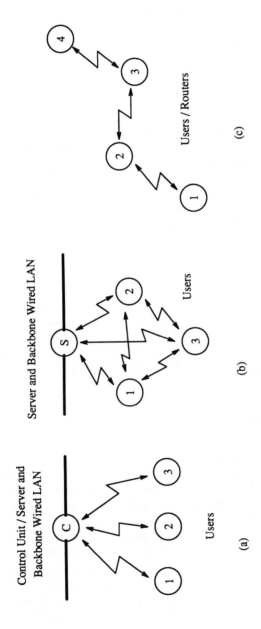

Figure 7.6 Three common topologies in wireless data communications: (a) central, (b) distributed, and (c) multihop.

Today, WLLANs are at the early stages of market penetration, and arge numbers of wireless terminals in offices are not expected in the near future. For limited numbers of users per office, the single failure point, the delay, and hertz-per-second usage characteristics are not as important as signal coverage and the ease of connection to the backbone network. As a result, the centralized topology is more popular than other topologies in today's market.

7.8.2 Distributed Architectures

In a distributed network, as shown in Figure 7.6(b), all terminals can communicate with one another directly. Communication with the backbone LAN is through a special server terminal. A distributed network does not suffer from additional store-and-forward delay, and the single transmission per packet minimizes the hertz-per-second usage for transmission of individual packets. Failure of any of the individual terminals cannot cause a failure of the entire network. However, a distributed architecture has a smaller coverage and is not suitable for the application of power control.

7.8.3 Multihop Architectures

A multihop topology, shown in Figure 7.6(c), has the best coverage among the all topologies. However, a multihop network requires a complex routing algorithm, has a high average hertz-per-second utilization and store-and-forward delay, is not suited for power control, and needs an additional terminal to bridge to the backbone wired network. Although the multihop architecture is popular for use in packet radio networks, it is not popular in the WLLAN industry.

7.9 MULTIPLE-ACCESS METHODS

Existing WLLANs are designed with data communication as their primary application. The multiple-access methods are different versions of ALOHA and carrier-sense multiple access (CSMA) rather than CDMA, TDMA, and frequency-division multiple access (FDMA) used in mobile and personal communication systems. In ALOHA, packets of data are broadcast as they arrive; if a collision occurs and the packet does not arrive at its destination, the packet is retransmitted. CSMA is the same as ALOHA except that the terminal senses the channel first, and if there is no other user broadcasting it broadcasts its packet. The advantage of these access methods is that the overhead is reasonable and the system can be designed and expanded easily. The problem is reduction in throughput caused by collision of the packets. In a nonfading wired network, the maximum throughput of the slotted ALOHA protocol is only 36%, and the maximum throughput of CSMA is slightly more than 50%. In radio channels, sometimes the differences in the level of received power from two packets involved in a collision is so large that one of the packets survives the collision, resulting in an increase in the throughput over the throughput in wired networks; this phenomenon is referred to as *capture*. Figure 7.7 illustrates the throughput of ALOHA and CSMA in Rayleigh fading and nonfading channels for various packet lengths and BPSK modulation [19,20]. Note that the differences among the signal power levels received from different terminals, if not controlled, cause a reduction of the bandwidth efficiency of the cellular TDMA, FDMA, and, in particular, CDMA systems.

In the existing WLLAN products, modifications are made to the original ALOHA and CSMA protocol to avoid collisions and further increase the throughput of the system. Packet reservation ALOHA, CSMA with collision avoidance, and persistent CSMA are the most popular access methods for WLLANs.

In a voice-dominated network, a secondary low-speed data channel can be established by taking advantage of the pauses between talkspurts. In a TDMA system with a large number of slots per frame, the unused slots can be utilized for data communications; an analysis of such a radio system is available in [21]. The same concept is applicable to FDMA networks by making use of idle channels in the existing FDMA analog cellular radio systems. In a CDMA network, integration is more natural. The data user can monitor

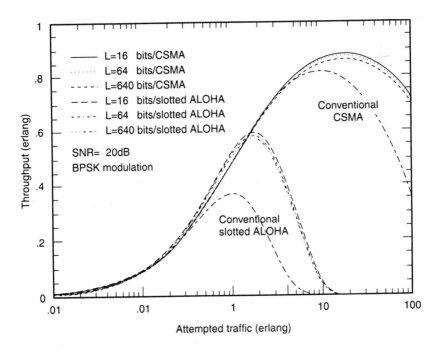

Figure 7.7 Throughput versus traffic performance of CSMA and ALOHA network and the effects of Rayleigh fading, modulation, and coding on the results.

the channel interference, and when it falls below a preset level, data packets can be transmitted through the system.

7.10 SUMMARY

Current wireless data communications are provided either by a high-speed WLLAN for local applications or by a low-speed mobile data service for wide-area coverage. The direction of the future is toward the integration of the two services in one battery-operated unit suitable for portable and personal computers, such as laptops and pen-pads. Development of this technology demands allocation of additional bands by the frequency administration agencies, further research and development in the low-power, multirate modems, and development of new systems by service providers.

ACKNOWLEDGMENTS

I would like to express my appreciation to Dr. Allen Levesque, for his careful editing and constructive comments, and Ganing Yang and Duan Wang, for their help in preparing the figures.

REFERENCES

[1] Devasirvatham, D. M.J ., "Time Delay Spread Measurements of Wideband Radio Systems Within a Building," *Electronics Letters*, Vol. 20, No. 8, 1984, pp. 949–950.
[2] Holt, T., K. Pahlavan, and J. F. Lee, "A Graphical Indoor Radio Channel Simulator Using 2-D Ray Tracing," Proc. 2nd IEEE Int. Symp. on Personal Indoor and Mobile Radio Comms. Boston, MA, Oct. 19–21, 1992, pp. 411–416.
[3] Saleh, A. M., and R. Valenzuela, "Statistical Model for Indoor Multipath Propagation," *IEEE J. of Selected Areas in Communication*, Vol. SAC-5, Feb. 1987, pp. 128–137.
[4] Ganesh, R., and K. Pahlavan, "On the Arrival of the Paths in Multipath Fading Indoor Radio Channels," *IEE Electronics Letters*, June 1989, pp 763–765.
[5] Rappaport, T. S., S. Y. Seidel, and K. Takamizawa, "Statistical Channel Impulse Response Models for Factory and Open Plan Building Radio Communication System Design," *IEEE Trans. on Communication*, May 1991, pp. 798–806.
[6] Ganesh, R., and K. Pahlavan, "Modeling of the Indoor Radio Channel," IRE Proc.-I, June 1991.
[7] Yegani, P., and C. D. McGillem, "A Statistical Model for the Factory Radio Channel," *IEEE Trans. on Communication*, Oct. 1991, pp. 1445–1454.
[8] Alexander, S. E., "Characterizing Buildings for Propagation at 900 MHz," *IEE Electronics Letters*, Sept. 1983, p. 860.
[9] Alexander, S. E., "Radio Propagation Within Buildings at 900 MHz," *IEE Electronics Letters*, Oct. 1982, pp. 913–914.
[10] Howard, S. J., and K. Pahlavan, "Autoregressive Modeling of Wideband Indoor Radio Propagation," *IEEE Trans. on Comm.*, Vol. 40, No. 9, 1992, pp. 1540–1552.
[11] Freeburg, T. A., "Enabling Technologies for Wireless In-Building Network Communications—Four Technical Challenges, Four Solutions," *IEE Comm. Mag.*, April 1991, pp. 58–64.

[12] Yang, G., and K. Pahlavan, "Comparative Performance Evaluation of Sector Antenna and DFE Systems in Indoor Radio Channels," *Proc. IEEE ICC*, Chicago, Vol. 3, 1992, pp. 1227–1231.

[13] Sexton, T. A., and K. Pahlavan, "Channel Modeling and Adaptive Equalization of Indoor Radio Channels," *IEEE J. of Selected Areas in Communications*, Vol. 7, No. 1, 1989, pp. 114–121.

[14] Pahlavan, K., S. Howard, and T. Sexton, "Adaptive Equalization of Indoor Radio Channel," *IEEE Trans. on Communications*, Jan. 1993, pp. 164–170.

[15] Acampora, A. S., and J. H. Winters, "A Wireless Network for Wide-Band Indoor Communications," *IEEE J. of Selected Areas in Communications*, June 1987, Vol. SAC-5, No. 5, 1987, pp. 796–805.

[16] Gfeller, F.R., and U. Bapst, "Wireless In-House Data Communication via Diffuse Infrared Radiation," IRE Proc. Vol. 67, No. 11, 1979, pp. 1474–1486.

[17] Yen, C. S., and R. D. Crawford, "The Use of Directed Beams in Wireless Computer Communications," *Proc. of IEEE Globecom*, New Orleans, Dec. 1985, pp. 1181–1184.

[18] Bary, J. R., et al., "Simulation of Multipath Impulse Response for Indoor Diffuse Optical Channel," *Proc. of IEEE Workshop on Wireless LANs*, Worcester, MA, May 1991, pp. 81–87.

[19] Zhang, K., and K. Pahlavan, "CSMA Local Radio Networks with BPSK Modulation in Rayleigh fading Channels," *IEE Electronics Letters*, Sep. 27, 1990, pp. 1655–1656.

[20] Zhang, K., and K. Pahlavan, "Relation Between Transmission and Throughput of the Slotted ALOHA Local Packet Radio Networks," *IEEE Trans. on Communications*, Vol. 40, No. 3, 1992, pp. 563–577.

[21] Zhang, K., and K. Pahlavan, "An Integrated Voice-Data System for Wireless Local Area Networks," *IEEE Trans. on Vehicular Technology*, April 1990.

Chapter 8
Spread-Spectrum Systems

Stephen Barton and Barry West

The term *spread spectrum* defines a class of digital radio systems in which the occupied bandwidth is considerably greater than the information rate.

The technique was initially proposed for military use, where the difficulties of detecting or jamming such a signal (the low probability of intercept, or LPI, and antijam properties) made it an attractive choice for covert communication. Only more recently has the civilian telecommunications community come to consider using spread-spectrum systems.

The aim of this chapter is to give the readers sufficient understanding of the principles of spread-spectrum systems and how they are implemented in practice, so as to enable them to make informed judgments of the validity of claims for particular offerings.

8.1 TECHNICAL OVERVIEW

Spread-spectrum systems are distinct from the other main class of wideband modulation systems based on nonlinear operations, which includes frequency modulation. The latter are able to trade excess bandwidth for improved tolerance to additive interference (*capture effect*) and noise, but they are characterized by threshold effects, with minimum signal-to-noise ratios below which they fail catastrophically.

Spread-spectrum systems, in contrast, are linear in the sense that the spreading operation in the transmitter can be completely reversed in the receiver, however low the signal-to-noise ratio in between. It follows that these systems offer no improvement in tolerance of noise in exchange for the excess bandwidth. They do, however, offer other benefits, including resistance to multipath propagation effects.

To spread the spectrum of an information-bearing binary stream, it is combined with a complex waveform having the desired spectral characteristics. This waveform may be:

- A series of pulses of carrier at different frequencies, in a predetermined pattern (*frequency hopping*, or FH);
- A pseudorandom modulating binary waveform whose symbol (or chip) rate is a large multiple of the bit rate of the original bit stream; this is known as direct sequence (DS) spread-spectrum.

The term *code-division multiple access* (CDMA) is often used in reference to spread-spectrum systems and refers to the possibility of transmitting several such signals in the same portion of spectrum by using distinct pseudorandom codes for each one. This can be achieved with either FH or DS spread spectrum and provides an alternative to frequency-division multiple-access (FDMA) or time-division multiple-access (TDMA) methods. However, the term CDMA has come to be used mainly in relation to DS spread-spectrum systems, which have attracted more attention than FH systems.

FH will be described briefly in the next section, the remainder of this chapter being devoted to a more detailed consideration of the theory and practice of CDMA based on direct-sequence techniques.

8.1.1 Frequency Hopping

In FH, the information is modulated onto a carrier using conventional narrowband modulation, and then the carrier frequency is shifted over the available bandwidth in a predetermined but pseudorandom sequence. If the chosen narrowband modulation is frequency-shift keying (FSK), then the information and hopping sequences can be combined digitally (least-significant and most-significant bits, respectively) before driving a common numerically controlled oscillator.

The LPI property comes from having a short dwell time on any particular frequency, while privacy is provided by keeping the hopping sequence secret. The antijam performance depends on the fraction of time spent on any particular frequency and, hence, the total number of frequencies available. A carrier wave (CW) jammer will only interfere with the signal for those hops when it falls on the frequency of the jammer. Interference from another FH signal similarly occurs for only this fraction of time, unless the two use the same sequence in perfect synchronism. Thus, FH-based CDMA is possible.

Multipath tolerance is based on the effects of frequency-selective fading. Narrowband signals can have large blocks of data obliterated when a spectral null falls on the carrier frequency. In FH, the carrier spends only a short time at each frequency, so that only a few bits are erased or received in error, whether by frequency-selective fading or collision with an interferer. This has a similar effect to time interleaving, that is, error bursts are separated and randomized so that the errors can be corrected by forward-error-correcting codes.

In a cellular system, individual transmitters in a ring of adjacent cells may be received at sufficient level to cause an erasure of the signal. However, the durations of these fades are limited to the dwell time of each frequency hop, so that the fade rate is controlled by the hopper. A much higher probability of erasure can be recovered by coding than is the case for narrowband systems, where a slow-moving user may exhibit fades of long duration.

8.1.2 Direct Sequence

Figure 8.1 represents the functions of a DS transmitter and receiver. the carrier is phase-shift key (PSK) modulated by the data at bit rate, R, and the spreading code at the much higher chip rate, W. In practice, these two modulators are usually combined, and the two bit streams combined in an exclusive OR gate. For the purpose of analysis, it is more convenient to treat them separately. In the receiver, the signal is modulated again by the identical spreading code in the despreader. Provided the codes are perfectly synchronized, the signal that appears at the output of the despreader will be the same as that at the input to the spreader. The signal energy, which was spread over W Hz in the channel, is thus collected into a bandwidth of only R Hz by the despreader. The spectral density of additive white noise is unaffected by this process. If the data demodulator requires a certain E_b/N_o to achieve the target bit error rate (BER), then the signal-to-noise power spectral density ratio in the channel is given by

$$E_c/N_o = (E_b/N_o)(R/W) \tag{8.1}$$

where E_c is the energy per chip.

For large bandwidth expansion ratios, the signal can be completely masked by noise at the receiver input. Thus, a casual receiver must be much closer to the transmitter to become aware of the signal, giving the LPI property. Privacy again results from keeping the spreading sequence secret.

Now consider a CW interfering signal, as shown in Figure 8.2(a). This signal passes through the despreader only, not the spreader. The result is that its energy is spread over the bandwidth W, and its energy density becomes $J_o = J/W$ W/Hz, where J is the power of the jamming, or interfering, signal. The fraction of this interference that falls in the band of the data demodulator, shown in Figure 8.2(b), is virtually indistinguishable from random noise and can be added to the thermal noise. Because CDMA systems are normally designed to be interference limited, we can, to a first approximation, ignore the thermal noise. Thus, the criterion for successful operation is

$$(E_b/J_o) = (S/R)/(J/W) \geq (E_b/N_o)_{min} \tag{8.2}$$

where $(E_b/N_o)_{min}$ is the minimum required for the design BER. The maximum jamming power, as a multiple of the signal power, is referred to as the jamming margin:

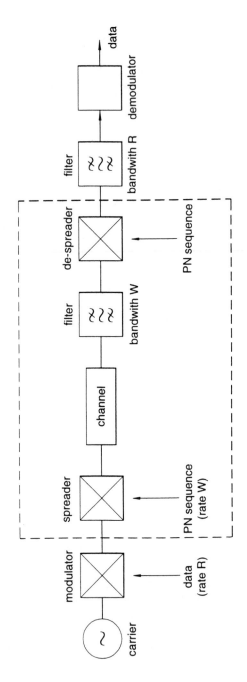

Figure 8.1 Direct-sequence spread-spectrum transmitter and receiver.

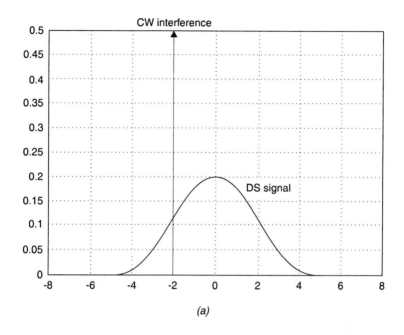

(a)

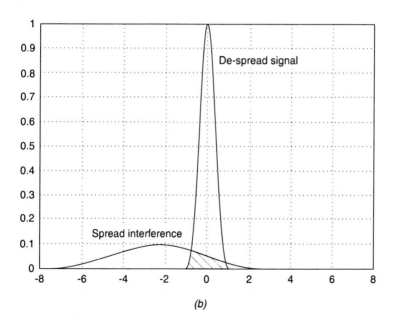

(b)

Figure 8.2 (a) Receiver input ($W = 5$). (b) Despreader output ($R = 1$).

$$J/S \le (W/R)/(E_b/N_o)_{min} \qquad (8.3)$$

The term W/R is referred to as the processing gain, G_p, of the system and is equal to the bandwidth expansion ratio.

Where the CW interfering signal is offset from the center frequency of the DS signal, as shown in Figure 8.2, the jamming power that falls in band will be reduced, so the margin will be increased, and (8.3) represents a worst case. Wideband interfering signals may be treated as a sum of many narrowband signals, and their contributions at the output of the despreader added. This corresponds to convolution of the interfering signal spectrum with the Fourier Transform of the spreading sequence. Again, (8.3) will be pessimistic. These comments apply to wideband interference, which is uncorrelated with the spreading code. They must be treated with caution in CDMA.

Multipath tolerance comes from the choice of codes that have good autocorrelation properties, that is, the autocorrelation falls to a very low value for all time offsets exceeding one chip. Then echoes arriving more than one chip period before or after the principal path are suppressed, reducing the delay spread seen by the data demodulator to no greater than the chip period, that is, much shorter than the bit period. Furthermore, echoes may be discriminated with a resolution of one chip period and received independently. This leads to the possibility of diversity gain in a Rake receiver structure. Thus, the DS system using a Rake receiver can actually turn the multipath channel to its advantage, achieving diversity gain. In contrast, the FH receiver merely affects the time durations of fades, leaving their probability distribution unchanged.

8.1.3 CDMA

In DS-CDMA, the various users of the system are allocated different spreading sequences or codes. The codes assigned to the different users are selected from families of binary sequences, each of which is noiselike (i.e., uncorrelated with itself shifted in time) and uncorrelated with other sequences in the family. (Here, *uncorrelated* must be taken to mean ''having an acceptably low correlation''). Suitable families of codes are the codes described as *Gold sequences*; an alternative approach is to use subsequences of a very long pseudorandom sequence. Since there may be a limited number of codes available, they are often assigned as required for the duration of a call, rather than being permanently allocated to each user. In addition, if transmissions are not to be synchronized within a system, a sequence must also be uncorrelated with the other sequences with any relative time shift. Conversely, if synchronization is to be applied, it is possible to use time-shifted versions of the *same* sequence for different users.

Consider the inbound (mobile to base station) link of a cellular radio system and assume initially that each of N mobile transmitters has its power controlled so that they are all received at the base with equal power. Then the interference power for any one user is the power of the other $N - 1$ users: $J = S(N - 1)$. Simple manipulation of (8.3) gives the minimum processing gain, or bandwidth expansion ratio:

$$(W/R) \geq (N - 1)(E_b/N_o)_{min} \qquad (8.4)$$

The spectral efficiency can be measured in terms of the bandwidth required per bit per second of total network traffic, that is,

$$(W/NR) \geq (E_b/N_o)_{min} \qquad (8.5)$$

If $(E_b/N_o)_{min} = 20$ (i.e., 13 dB) for example, the bandwidth requirement for this simple single-cell model would be 20 times the aggregate data rates. This type of calculation led to the rejection of CDMA as a candidate for such systems for many years. However, the bandwidth requirements can be reduced considerably by forward error correction (FEC) coding and voice activation (VA), and the extra bandwidth required in a multicell environment is much less than is the case for conventional systems.

8.1.4 FEC and VA

In conventional systems, FEC provides a reduction in the $(E_b/N_o)_{min}$ requirement in exchange for an increase in occupied bandwidth. In CDMA, the reduction in $(E_b/N_o)_{min}$ translates into a reduction in occupied bandwidth. There is no penalty, other than hardware complexity, for the use of very powerful, very low rate codes. As an example, if coding reduces the $(E_b/N_o)_{min}$ requirement to 6 dB, then the bandwidth required per bit per second of traffic is reduced to 4. The only limit to this improvement is set when the code rate falls to R/W. Beyond this point, the bandwidth requirement will be set by the code rate rather than the jamming margin. Codes with rates less than 1/4 are generally so complex as to be impractical, so this limit is not likely to be approached.

In conventional systems, VA offers a saving in transmitter power, but this is not readily translated into a bandwidth saving. The average voice activity ratio in telephone calls is about 40%. This means that in a system with 100 calls in progress the number of callers actually speaking rises above 50 only 2% of the time. Thus, the interference level can be taken as half of the worst-case value, reducing the bandwidth requirement by a further factor of 2.

The spectral efficiency of CDMA with FEC and VA in the single-cell scenario is thus comparable to that of conventional FDMA or TDMA, that is, about 0.5 bit/s/Hz. The major advantage over the conventional systems becomes apparent when we consider the multiple-cell configuration.

8.2 SYSTEM IMPLEMENTATION

Having reviewed the fundamentals of spread-spectrum systems, we will now deal with the more detailed design aspects that have to be considered in implementing a practical system. Although CDMA is a fairly simple idea in principle, considerable complications arise in practice to realize the performance potential of these systems. The main issues ar discussed in this section; however, real CDMA systems include yet further complication introduced with the object of optimizing performance.

8.2.1 Multiple-Cell Configuration

In conventional cellular systems, different bands of frequencies must be allocated i neighboring cells. With a cluster size of 7, for example, the total system bandwidt required for N users per cell is seven times that required in the single-cell case.

In the CDMA case, the entire available bandwidth is normally used in every cel Users in neighboring cells contribute to the total interference power received at the bas station. The advantage over conventional systems lies in the fact that the interference i made up of the sum of a large number of signals. The variance of the total interferenc is therefore small, and capacity calculations can be based on the mean power. In contras in conventional systems an outage occurs if a single user allocated the same frequenc in a distant cell is received with a power level of only one-tenth the wanted signal. Thus the probability distributions of individual signal powers must be considered, out to th 99% to 99.9% probability range.

Figure 8.3(a) shows the transmitter power versus normalized distance of the transmit ter from the base station for a system with perfect power control and inverse 3.5 powe law propagation. The worst- and best-case directions, A and B, are shown in Figure 8.4 Figure 8.3(b) shows the received power at the base station as a function of the normalize distance. The flat part from 0 to 1 corresponds to the "own-cell" users, which are power controlled by the base station.

The remainder of the curve corresponds to users who are power-controlled by othe base stations. It is clear that the contribution to the total interference power received from users outside the cell is small compared to the own-cell interference. Detailed simulation in [1] show that this contribution increases the interference by only about 50%. Thus, th bandwidth required per bit per second per cell in the CDMA case is only 1.5 times th figure for a single cell. This should be compared with the factor of 7 for conventiona systems with a seven-cell cluster.

The plots of Figure 8.3 are based only on the power law propagation, and ignor shadowing. Since the interfering signals add on a power basis, the means and variance of their power distributions must be added. With perfect power control, own-cell interferer have mean S and zero variance. The sum of many adjacent cell interferers with log manual distributions will of course be Gaussian with a relatively narrow standard deviatior Figure 8.5 shows a log-normal distribution converted to a linear scale of power. Becaus of the long upper tail, the mean power is always greater than the median. Figure 8. shows how this increases with the log-normal standard deviation. This implies that wit standard deviation as high as 12dB (8 dB on each path) the "out-of-cell" interferenc could be up to 27 times that calculated on the basis of the power law alone. Howeve

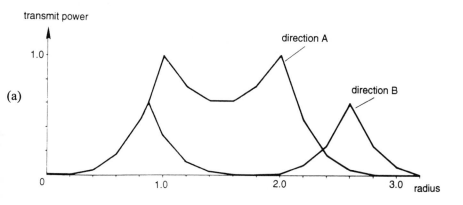

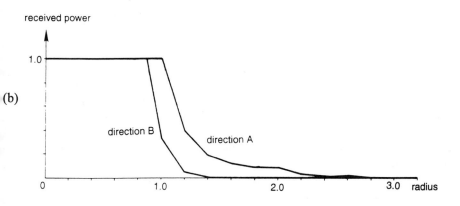

Figure 8.3 (a) Power control, $r^{-3.5}$ law. (b) Received power versus radius from base station.

This is based on an implicit assumption that every user is allocated to the geographically nearest base station rather than the one with the lowest path loss and ignores handoff.

8.2.2 Handoff

In practice, the spectral efficiency is maximized by controlling every user to transmit the minimum power necessary to enable one base station to demodulate its signal with the required BER, regardless of which base station that is. This requires all base stations to monitor users in neighboring cells and capture any that is received with lower than the target BER, so as to command it to reduce power. If this could be achieved, no interfering signal would ever be received with power greater than S. This reduces the mean and the variance of the interfering signal power. It is claimed in [1] that this results in out-of-cell

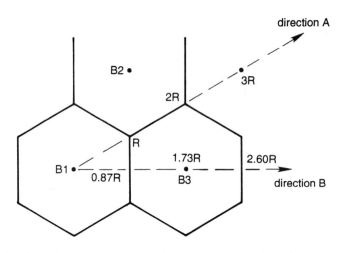

Figure 8.4 Geometry for Figure 8.3.

interference power of only 67% of the own-cell interference, which leads to a spectral efficiency of 0.3 bit/s/Hz in every cell.

In practice, the handoff procedure will not be perfect, for a number of reasons. Resources committed to monitoring signals in neighboring cells must be limited so that some sort of polling will be required. Reacquisition times will set minimum intervals between sampling interfering signals. The accuracy with which the BER can be measured will be limited by the dwell time on each user and the rate of change of its signal strength. Hysteresis must be introduced to avoid rapidly repeating handoffs and consequent switching of traffic in the fixed network. All these factors lead to higher levels of interference power and, hence, bandwidth required for the CDMA system.

8.2.3 Power Control

In any mobile radio system, transmitted power (in both directions) may be adjusted in response to information sent from the other end about the strength of the received signal, to a level that is just sufficient to ensure adequate transmission quality.

In FDMA and TDMA systems, interference between users in the same cell can be reduced as far as is desired by appropriate filtering and by suppression of transmissions other than in one's own time slot. Power control is, however, often applied to improve spectral reuse in other cells and to prolong battery life in portable equipment.

In CDMA systems, imperfect power control affects the interfering-to-wanted signal power. Suppose the power control reduces the dynamic range of signal powers to a residual ±3 dB. This marginally increases the total interfering power and standard deviation, but the effect is quite small. On the other hand, if the wanted signal can be 3 dB below its nominal level, the jamming power-to-signal ratio is increased by a factor of 2 for this

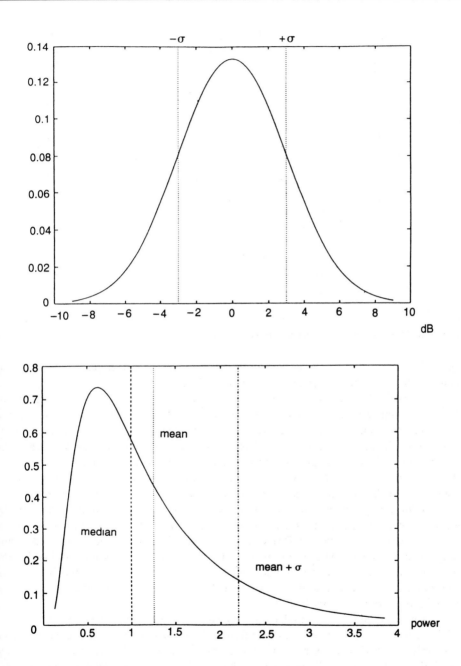

Figure 8.5 Log-normal distribution, $\sigma = 3$ dB.

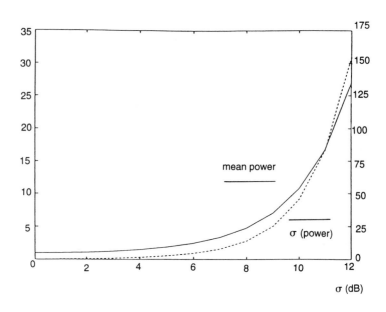

Figure 8.6 Mean and standard deviation of power distribution.

user. Combining the effects of handoff hysteresis and imperfect power control, it is unlikely that a spectral efficiency greater than 0.2 bit/s/Hz can be achieved.

A limited dynamic range in the power control circuitry can also lead to problems. Consider a mobile user very close to the base station. The power control loop must cause the mobile to reduce its transmit power so that it is received at the same level as all the other users. If the transmitter power has a minimum value below which it cannot be reduced, this will generate proportionately higher interference to the other users. This is the so-called *near-far problem*, which in practice means that fast-acting, wide-range power control has to be applied if the maximum capacity is to be realized. Another way of overcoming this problem is to use a base station antenna with a cosec[4] pattern in the elevation plane. It is normal practice to apply such a law as a minimum gain to ensure there are no deep nulls, but for CDMA a maximum gain is also required. Then the dynamic range requirements of the user terminal is only the range of the log-normal distribution, plus the difference between minimum and maximum antenna gains. This strategy will cause all users to transmit at the same nominal power level, and this will also affect the out-of-cell interference contribution.

When the number of simultaneous talkspurts exceeds the design figure, the BER of all users will increase above the target figure. Under the power control regime described here, this would cause all users to increase their transmit power, but this would not reduce the BER. An unstable condition arises in which the interference power generated by users in the overloaded cell grows without limit. This condition must be recognized and prevented from becoming unstable by modifications to the power control algorithm.

8.2.4 Diversity

Diversity encompasses a number of techniques used to counteract the fading behavior of the mobile radio signal: some of these are applicable to TDMA and FDMA as well as CDMA systems.

- Space (or path) diversity using multiple antennas at the fixed sites, to obtain two or more versions of the signal that are subject to independent fading and that are combined to form a resultant signal with better statistical properties.
- Multipath diversity, whereby versions of the signal arriving via separate paths (and thus at different times) are optimally combined at the receiver.
- Macrodiversity, that is, simultaneous use of links between the mobile and two or more fixed stations. This technique obviously introduces inefficiencies in the use of equipment and also in the reuse of spectrum. However, it can be used temporarily to provide a smooth transition as the mobile hands off from one base station to another (soft handover).
- Frequency diversity, relying on the frequency-selective nature of the multipath fading. For this technique to work, it is necessary that the (spread) signal bandwidth be at least comparable with the coherence bandwidth of the channel (see Chapter 3).
- Time diversity, by using symbol interleaving and error-correcting coding to introduce time correlations into the signal.

8.2.5 Synchronization in Multipath

When a transmission is first initiated, the receiver has to "lock on" to the incoming signal, which will have been subject to propagation delays and other effects; a synchronization procedure is therefore needed. The problem is complicated by the multipath propagation found in mobile radio channels with typical delay spread of, say, 20 μs, which may well exceed the chip period.

The usual solution is the Rake receiver. This correlates the received signal with multiple copies of the code, each with a different delay offset. The correlator outputs are then independently fading copies of the same message and can be combined in a maximal-ratio combiner to achieve, in effect, a form of multipath diversity.

8.2.6 Hybrid Systems

CDMA systems will frequently divide the available bandwidth among several radio carriers, each modulated with an ensemble of CDMA user signals, thus forming a hybrid CDMA-FDMA system.

Similarly, systems have been proposed that use a radio carrier bearing a number of code-divided data streams, which are each also time-divided between several users, that is, a CDMA-TDMA hybrid. Finally, it is possible to envisage using a combination of all three multiple-access techniques in a system.

8.2.7 Capacity (Spectral Efficiency)

Several authors have presented capacity estimates for CDMA systems, based on analytical, simulation, and practical results. In [1], an analysis based on certain idealizing assumptions gave an estimate of 36 users/MHz/cell. Later authors [2–6] have introduced refinements of this analysis, taking into account various implementation aspects; these give capacities ranging from 15 to 39 users/MHz/cell. Later field trials appear to verify that capacities within this range are attainable in practice.

It is of interest to compare these estimates with a TDMA system. The capacity of the GSM cellular system, assuming 7 users per 200-kHz carrier and 7-cell frequency reuse, is 5 users/MHz/cell; with half-rate speech coding, permitting 13 users/carrier, and with 4-cell reuse, this would increase to 16.3 users/MHz/cell.

An advantage of CDMA is that it exhibits *soft capacity*; whereas a fully loaded TDMA system will be unable to cope with an additional call, a CDMA system may trade capacity for quality. Thus, if it is desired to exceed design capacity in a particular cell, for example, temporarily to allow a handoff to take place, this can be done at the cost of slight quality degradation for all users.

8.2.8 Coexistence With Other Systems

As explained in the introductory paragraphs of this chapter, the military user of CDMA is attracted by its LPI and privacy and by its antijam characteristics—an enemy will not be able to detect the signal and will also find it difficult to cause interference to the CDMA system. It is tempting to deduce that spread spectrum will not interfere with (or be interfered with by) a communication system using conventional modulation, since the two systems will not ''see'' each other, except as a slight increase in the background noise level. Therefore, the argument goes, CDMA systems can reuse spectrum already occupied by, for example, microwave line-of-sight fixed links [7].

This claimed advantage of CDMA is, however, controversial, and the matter is more complex than might appear. In practice, interference effects have to be evaluated in any given case, often by use of computer simulation. The qualitative discussion in this subsection explains the main factors to be considered in such an evaluation.

In a CDMA system, the power transmitted in respect of a single user can be spread over the available bandwidth so that the spectral density (power per Hertz) is indeed below the level of thermal noise, at the (intended) receiver. However:

1. A CDMA system normally operates in an interference-limited mode, in order to maximize the amount of traffic in a given bandwidth. Transmitted powers will

therefore be higher than in a conventional system to overcome this background of interference. Further, the level of interference to the other system will be the summation of transmitted output for all the users in the cell.

2. It is true that various techniques are used in CDMA systems to reduce the transmitted power (e.g., fast power control); however, these techniques are intended to increase the total number of users, and this offsets the apparent advantage.

3. Because of the relative positioning and directional properties of intended and unintended receiver antennas or because of propagation effects, the interference effect may happen to be greatly increased in a specific instance; the assessment must thus be carried out for this worst case.

In the particular case where the "other" system is a fixed microwave link, further considerations apply:

4. Fixed links are usually designed with very large signal-to-noise margins, in order to provide a very high availability, of perhaps 99.99%. This does not mean that the noise floor may safely be increased by allowing interference, since this will erode the ability of the link to cope with the (rare) propagation events causing large reductions in signal level.

5. The CDMA system will be designed on the basis of propagation laws appropriate to, say, urban conditions (e.g., Hata model; see Chapter 3). It must, however, be remembered that the link antennas will be placed clear of obstructions, perhaps on a hilltop or a tower; propagation between it and a mobile CDMA user will often be line of sight (i.e., square-law), and account should be taken of this difference in the statistical analysis of interference effects.

In one reported study of a particular scenario [8], it was concluded that, although there was some inherent protection due to the different modulation methods, microwave links would nonetheless have to be protected from the CDMA system by *exclusion zones*. When a mobile approaches such a zone, it must hand over, either to an alternative CDMA carrier (if there is more than one carrier available) or to a different air-interface; neither solution is attractive to the system implementor.

Conversely, the CDMA system may sometimes require to be protected from nearby link transmitters, for example, by the use of notch filters.

The overall conclusion is that it is not in practice possible to plan and regulate CDMA systems independently from other systems sharing the band.

8.3 PRACTICAL DEVELOPMENTS

8.3.1 Experimental Networks in the United States

A substantial amount of effort has been devoted by U.S. interests toward the development, trialling, and promotion of various alternative approaches. The main players have been:

- Radio equipment manufacturers (e.g., Qualcomm and Cylink);
- Cable television companies (e.g., Time Warner, Comcast, Cox Enterprises);
- Other carriers, including the phone companies.

The regulatory framework in the United States, determined by the FCC, has generally allowed great flexibility for the proponents of various approaches (including CDMA) to mount experimental trials (see Chapter 9).

Some system trials have been based on the "Part 15" regulations, using the industrial, scientific and medical (ISM) bands: 902 to 928 MHz, 2400 to 2483.5 MHz, and 5725 to 5850 MHz. Other experimenters have used the 1850 to 1990 MHz band, which is within the part of the spectrum identified for European and global allocations for PCN systems. Yet other trials (including some by Qualcomm) have been conducted using Short Term Authorization and Development Authorization license procedures for replacing FDMA in existing cellular bands.

Overall, the situation in the United States looks chaotic; the final outcome may depend as much on commercial considerations as on the relative technical merits of the various approaches.

8.3.2 European Initiatives

Development of CDMA systems in Europe is considerably behind that in the United States, but there is a growing tide of interest, manifested in various activities:

- The RACE initiative (R&D in Advanced Communications Technologies in Europe), associated with the European Commission Third Framework Program, includes several projects on mobile communication and one in particular (CODIT) on CDMA, which will construct and trial an experimental CDMA testbed.
- COST-231, "Evolution of Mobile Radio Communications," is a forum for interchange of research results and has included some work on CDMA systems, including study of various trials results, and comparisons with alternative access techniques.
- The European Telecommunications Standards Institute (ETSI) was responsible through its SMG Technical Committee for the GSM standards; the same group has created a subcommittee (SMG-5) to study approaches, including CDMA, to third-generation systems.
- National initiatives, for example, the United Kingdom has a partly government-funded project within its LINK Personal Communications Program, involving a number of companies and universities, which will produce and trial an experimental CDMA system aimed at PCN applications.

8.4 CONCLUSIONS

In recent years, spread spectrum has come from a narrow range of military applications to be a major contender in large-scale civil applications such as cellular radio and multibeam atellites. Many of the advantages attributed to CDMA have not been fully verified, and here is in particular a lively debate going on about whether the claimed spectral efficiency ind low transmitter power requirements will be achieved in practice. The questions can only be resolved by careful analysis of the results of practical trials.

REFERENCES

[1] Gilhousen, K. S., et al., "On the Capacity of a Cellular CDMA System," *IEEE Trans. on Vehicular Technology*, Vol. VT-40, No. 2, May 1991, pp. 303–312.

[2] Gudmundson, B., et al., "A Comparison of CDMA and TDMA Systems," *Proc. 42nd IEEE Vehicular Technology Conf.*, Denver, May 1992, pp. 732–735.

[3] Chebaro, T., and P. Godlewski, "About the CDMA Capacity Derivation," *URSI ISSSE '92 Symposium*, Paris, Sept. 1992.

[4] Hulbert, A. P., and R. J. Goodwin, "Factors Affecting Spectral Efficiency in a CDMA Cellular Network," *IEE Colloq. on Spread-Spectrum Techniques for Radio Communication Systems*, London, June 1992.

[5] Newson, P., "The Effect of Power Control Error on the Capacity of a Direct Sequence CDMA System for Mobile Radio," *IEE Colloq. on Spread-Spectrum Techniques for Radio Communication Systems*, London, June 1992.

[6] Stuber, G. L., and C. Kchao, "Analysis of a Multiple-Cell, Direct-Sequence CDMA Cellular Mobile Radio System," *IEEE J. of Selected Areas in Communications*, Vol. 10 No. 4, May 1992, pp. 669–679.

[7] Schilling, D. l., et al., "Spread Spectrum for Commercial Communications," *IEEE Comms Mag.*, April 1991.

[8] "The Microcell Report," *J. Microcell Telecomms.*, Vol. 2, No. 4, April 1991.

Chapter 9
The Role of Satellites in PCS

John Gardiner

The motivation for incorporating satellite-delivered services in a future systems architecture is readily perceived: the emphasis that has been placed on adaptability in terminal-operating characteristics derived from the expectation that individual terminals will reconfigure their air-interface parameters to take advantage of a wide range of differing services. These will range from high-bit-rate integrated services digital network/integrated broadband communications network (ISDN/IBCN) transparent services in wireless office environments to low-bit-rate data- or voice-oriented services in rural cellular environments.

As has been apparent in trends in technical developments and in the spectrum-regulatory processes that have defined radio resources to support anticipated demand, operating frequencies for personal terminals are being identified at progressively higher regions of the spectrum. The future public land mobile telecommunication system (FPLMTS) has been allocated spectrum in the region of 2 GHz; spread-spectrum systems, anticipating the use of ISM bands are targeting the 5- and 17-GHz bands and so on. The impact of migration to higher frequencies on systems architecture is generally to increase the number of base stations required to achieve service availability over large areas and to make it progressively less attractive in economic terms to provide coverage in rural areas with low user demand. It is in this context that the primary appeal of satellite PCS is based, since the capability of satellites to provide wide-area coverage, albeit with limited user capacity, has been perceived as an answer to the problem of rural service provision. However, other possibilities have also been explored as the concepts of layered systems have been developed with accompanying definition of techniques for overlaying macrocells on microcells, which may in turn be overlaid on picocells. In such system architectures, the satellite macrocell might be regarded as a standby resource available to terminals anywhere if terrestrial based cells are congested.

This latter service objective has, however, considerable impact on the design and configuration of the satellite system component since, by implication, access to a satellite overflow resource requires the spacecraft to be visible from the ground in a variety c typical operating circumstances from open motorways to urban environments. Thes considerations have resulted in a wide variety of differing approaches to the design o the space segment in third-generation PCS; this last part of the forward vision of futur systems presents an overview of these approaches and discusses their advantages an disadvantages with prospects for implementation.

9.1 ORBIT OPTIONS

A primary factor in providing service to handportable units is link budget, and allied tc this is the question of operating frequency. WARC '92 has earmarked allocations in L band and S-band for satellite up- and down-links to portable terminals (1980 to 201(MHz and 2170 to 2200 MHz), frequencies at which minimal problems with rain attenuatior can be expected. This has prompted investigation of the feasibility of delivering servic(to handportables from orbits that place the spacecraft at distances of up to 39,000 kn from Earth. However, while it is possible, at least in principle, to deliver adequate flu> density on the ground from spacecraft with powerful on-board amplifiers and large high gain antennas, the return path is restricted in capability by the levels of radiation from handportables that can safely be used and the available DC power from handportable batteries.

A further factor in the debate about how to integrate space and terrestrial segments is the delay introduced by the length of the radio path between terminals on the ground and the spacecraft.

The third most significant parameter is spacecraft visibility, bearing in mind the possibilities of providing service in areas where buildings or geographical features may restrict the arc of sky visible at any one time.

9.1.1 The Geostationary Orbit (GEO)

This option, the basis of the great majority of satellite-based telecommunications services, has also been the space resource that has supported the now well-established satellite mobile services to ships at sea, operated by the International Maritime Satellite Organization (INMARSAT). In the maritime application, however, there is no requirement for operation to handportable terminals. Installations on ships can use antennas with significant gain, as can, in extending operation to land mobile terminals, the ground station terminals used for electronic news gathering. High-gain antennas are highly directional, so that antenna steering, automatically on board ship and manually for ENG terminals is an essential feature. Operation to handportable terminals without the advantage of directionality and gain in the antenna results in somewhat limited services—paging and low-bit-rate data. The advantages of geostationary service provision are, however, that spacecraft resources are available, as is the infrastructure of ground Earth stations, network control facilities, nd so on. Moreover, a single satellite is adequate to provide 24-hour service. However, ignificant disadvantages remain in long round-trip propagation delays and in restricted visibility of geostationary satellites in northern latitudes. A geostationary spacecraft is only some 22 degrees above the horizon in much of United Kingdom, for instance.

.1.2 The Highly Elliptic Orbit (HEO)

An alternative to the GEO approach, which has been developed extensively in the former Soviet Union, utilizes the highly elliptic orbit depicted in Figure 9.1. This orbit, inclined at 63.4 degrees positions the spacecraft at apogee near to zenith at the appropriate longitude. This means that the satellite service is accessible to users on the ground even in densely built-up urban areas. The disadvantages are that the apogee of the orbit is as far from the Earth's surface as is the geostationary orbit, so that link budgets and delay problems are similar with the added complication that the satellite is now in motion relative to the Earth. This means that there is significant and variable Doppler shift to be compensated for and that the round-trip delay also changes as the spacecraft approaches apogee and then returns from it. Also, the window during which the spacecraft is overhead lasts for around 8 hours, so that to obtain 24-hour coverage requires that three satellites be operational and that appropriate handover strategies can be implemented to enable one satellite

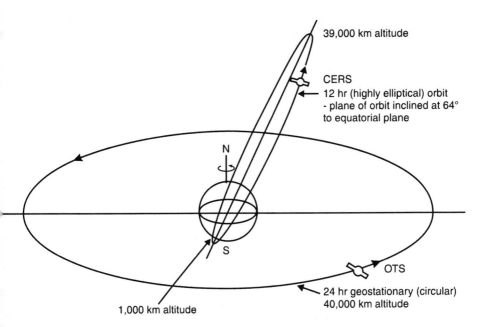

Figure 9.1 Highly elliptical orbit (HEO).

to take over service provision as it enters the active part of the orbit and the departing satellite leaves.

Despite these disadvantages, however, this approach has attracted a substantial amount of research effort, in the United Kingdom within the Communications Engineering Research Satellite (CERS) Project and subsequently the Technology Satellite (T-SAT) Project [1]. This research has subsequently been pursued by the European Space Agency under their ARCHIMEDES program [2].

9.1.3 Circular Orbits

If it is accepted that a number of satellites may be required to provide the desired services then further alternative strategies are possible utilizing low earth orbit (LEO) constellations. The lower the altitude of the satellite, the shorter will be the orbit period, and the more satellites will be needed to sustain continuous coverage on the ground. Table 9.1 shows a typical tabulation of the relationship between the number of satellites required and the angle about vertical within which at least one satellite is visible at any one time. Naturally since the Earth rotates underneath the orbiting spacecraft, the constellations need to be configured in orbit planes with a number of satellites assigned to each orbit plane.

Table 9.1
Circular Orbit Configurations

h (km)	El°	N
1000	5	41
1000	10	57
1000	15	79
1200	5	35
1200	10	47
1200	15	63
1400	5	30
1400	10	40
1400	15	52
1600	5	26
1600	10	35
1600	15	45
1800	5	24
1800	10	31
1800	15	40
2000	5	22
2000	10	28
2000	15	35

h = altitude of satellites; $El°$ = angle above which satellites are visible; N = total number of satellites needed.

It is easy to see that if the orbits are configured to pass over the poles, then many spacecraft are visible in polar regions but fewer in equatorial areas. An alternative to polar orbits is provided by the Walker orbit [3], which, while circular, is inclined relative to Earth's rotational axis, as shown in Figure 9.2. It is seen that this arrangement has the effect of increasing the number of visible satellites within a given vertical solid angle on Earth's surface in the equatorial region at the expense of the polar coverage. The choice of orbit altitude is not unconstrained, however, due to the presence of the Van-Allan radiation belts, which encircle Earth and which represent a hostile environment for spacecraft electronics. The densest proton radiation belts are centered at about 3000 and 7500 km above Earth's surface, and operation of active satellites within these bands is not feasible with current or envisaged on-board electronics, particularly on-board processing.

Two alternatives are therefore possible: to station spacecraft in orbits either below about 1500 km or above about 10,000 km, so as to avoid the major radiation belts. One strategy results in the so-called LEO constellations, the other in medium Earth orbit (MEO) constellations. The attractions of the LEO/MEO approaches are readily appreciated. First, link budgets are greatly improved, making handportable service provision much more readily achievable. Round-trip delay is also greatly reduced, and while some Doppler shift results from the passage of satellites over the terrain beneath, this is not so pronounced as that resulting from HEO implementation.

Figure 9.2 "Walker" orbit.

There are disadvantages, however. The number of satellites required in LEO constel lations to support service becomes large, and the more spacecraft that are involved, the more complicated becomes the handoff procedures needed to transfer traffic from one satellite to another as they pass overhead. There is a further complication: LEO and MEO constellations inherently provide global coverage—the implications of this coverage from the standpoint of regulatory and licensing procedures pose significant problems for the international radio regulatory community [4].

9.2 PROPOSALS FOR SATELLITE PERSONAL COMMUNICATIONS NETWORKS (S-PCN)

As part of the process of determining the need or otherwise for satellite personal communi cations network (S-PCN) standards in Europe, it is clearly important to undertake a review of the current proposals that have been made by companies and other organizations for such S-PCN systems.

In discussion of these proposals, it has become conventional to talk of "little LEOs" and "big LEOs" as a means of discriminating between classes of system and the kinds of services they provide. Little LEOs do not concern us in the context of PCS proper since they are assigned to frequencies below 1 GHz and offer only restricted data and positioning services. Big LEOs, on the other hand, are designed to offer speed services in the WARC '92 bands as a minimum requirement.

9.2.1 The Aries System

The system called Aries, proposed by Constellation Communications Inc., is being designed to provide mobile voice, data communications, and positioning through a constel lation of 48 lightweight (200 kg), transparent satellites in circular polar orbits providing seamless global coverage. Communications links are established between the user terminal and a gateway interconnected to the public-switched telephone network (PSTN) and other gateways via the terrestrial network. Two types of user terminals are planned, a portable but not hand-held unit and a mobile unit to be mounted on various mobiles. The frequency bands and bandwidths are given in Tables 9.2(a) and (b).

9.2.1.1 Air-Interface

The access scheme is TDM/SDMA (direct sequence spreading over 16 MHz, 3/4 rate FEC Viterbi soft decision decoding, K = 7 in the forward direction and FDMA/SCPC QPSK (FEC as in the forward link) with 50% filtering in the return direction. The terminal has memory-stored codes to acquire a CDMA carrier at switch on.

The channel rate for 4.8 kbit/s vocoded voice is 6.4 kbit/s.

Table 9.2(a)
Aries Frequency Bands

	Band
Service uplinks	1624.5 to 1626.5 MHz
Service downlinks	2483.5 to 2500 MHz
Feeder uplinks	C (6.5 GHz)
Feeder downlinks	C (5.1, 5.2 GHz)

Table 9.2(b)
Aries Constellation

Orbit	Altitude (km)	Period (hr)	Inclination	Number of Planes	Plane Spacing (deg.)	Satellites per Plane
Polar Circular	1020	1.75	90	4	45	12

9.2.1.2 Data (Real-Time/Store–and–Forward)

Data and facsimile transmission services are proposed in both directions as part of the same basic network architecture that provides telephony. Data collection, distribution, and control services are provided on a polled basis using two-way channels or on a random basis using packet messaging over a channel configured for signaling operations. The data rate is 2.4 kbit/s with a bit error rate (BER) of 10^{-5}.

9.2.2 Diamond (British Aerospace Satellite Systems Ltd.)

Diamond (see Table 9.3) is an S-PCN system utilizing HEOs and is proposed by British Aerospace Satellite Systems Ltd. of the United Kingdom. The proposals for Diamond are still relatively new, and much remains undefined; the following description is, therefore, somewhat brief.

9.2.2.1 Frequency Bands and Bandwidths

The frequency bands required for the Diamond system are as shown in Table 9.3(a), and the bandwidth needs are summarized in Table 9.3(b).

Table 9.3(a)
Diamond Frequency Bands

	Band
Service uplinks	L-band
Service downlinks	L-band
Feeder uplinks	Ku-band
Feeder downlinks	Ku-band
Intersatellite links	Optical, if needed

Table 9.3(b)
Diamond Bandwidth Requirements

	Bandwidth Requirement (MHz)
Service uplinks	25
Service downlinks	35
Feeder uplinks	200
Feeder downlinks	200
Intersatellite links	Not applicable

9.2.2.2 Orbits

Six satellites are used in a 48-hour repeat HEO to provide continuous coverage. Three close launches of satellite pairs provide for a full constellation; as an alternative, there is an initial deployment of three satellites in a Molniya 8-hour orbit, which are then boosted to a 48-hour orbit after a second deployment of three satellites. An initial configuration of three satellites in an 8-hour HEO would provide coverage of only one region, however.

The configuration of the constellation ensures that there is always at least one satellite visible to a user or a gateway at a minimum elevation angles for a usable service of approximately 55 degrees.

9.2.3 Ellipso (Ellipsat Corporation)

The Ellipso system, proposed by Ellipsat Corporation, is a satellite-based network intended to extend and complement existing commercial terrestrial mobile communications services. It is conceived as a global system and provides voice, real-time, and store and forward data, positioning and position reporting, and paging services.

Ellipsat Corporation emphasizes that, at the time of writing, the system design is tentative and subject to change pending the outcome of the U.S. negotiated rulemaking procedures, which relate particularly to spectrum allocation and sharing issues and possible standardization.

9.2.3.1 Frequency Bands

The frequency bands and bandwidths required for the Ellipso system are detailed in Tables 9.4(a) and (b).

9.2.3.2 Orbits

The Ellipso system uses two classes of orbits: inclined elliptical sun-synchronous orbits and equatorial circular orbits. The constellation is defined in Table 9.4(c).

The constellation is designed to cover latitudes of the Earth in proportion to population by latitude. An additional feature is the adjustment of the inclined elliptical orbits arguments of perigee and right ascensions of the ascending nodes to favor a daytime service.

The Ellipso constellations are designed to offer flexibility in deployment. Marketable services will be introduced in phases, corresponding to deployment level; deployment is targeted to begin in 1995. Ellipso satellites will be launched up to five at a time to build

Table 9.4(a)
Ellipso Frequency Bands

	Band (MHz)
Service uplinks	1610 to 1626.5
Service downlinks	2483.5 to 2500
Feeder uplinks	6452 to 6725
Feeder downlinks	5150 to 5216
Intersatellite links	None

Table 9.4(b)
Ellipso Bandwidth Requirements

	Bandwidth Requirement (MHz)
Service uplinks	16.5
Service downlinks	16.5
Feeder uplinks	273
Feeder downlinks	66
Intersatellite links	Not applicable

Table 9.4(c)
Ellipso Constellation

Orbit	Apogee (km)	Perigee (km)	Period (hr)	Inclination (deg)	Ascending Node	Argument of Perigee	Number of Planes	Satellites Per Plane
Inclined elliptical	7800	520	3	116.5	Under review	−270°	3	5
Equatorial circular	8068	8068	4.8	0	n/a	n/a	1	9

the constellation. Replenishment will use smaller boosters to launch satellites one or two at a time.

9.2.3.3 Terminal Characteristics

The Ellipso user terminals are comparable to those used for terrestrial cellular telephony. Ellipso is designed to facilitate seamless switchovers from terrestrial cellular to Ellipso communications when cellular becomes unavailable to the user.

Ellipso supports two types of user terminals:

- Mobile terminals, designed for vehicular use and capable of 5W or more of RF power;
- Handheld terminals, limited to a maximum of 1W.

9.2.4 Globalstar (Loral Qualcomm)

The Globalstar system, proposed by Loral Qualcomm, is a CDMA-based transparent satellite network designed to integrate deeply into terrestrial public land mobile netoworks (PLMNs) to provide voice, data, and positioning services. The frequency bands and bandwidth required for the Globalstar system are given in Tables 9.5(a) and (b).

Table 9.5(a)
Globalstar Frequency Bands

	Band (MHz)
Service uplinks	1610.0 to 1626.5
Service downlinks	2483.5 to 2500
Feeder uplinks	6484.0 to 6541.5
Feeder downlinks	5158.5 to 5216.0
Intersatellite links	None

Table 9.5(b)
Globalstar Bandwidth Requirements

	Bandwidth Requirement (MHz)
Service uplinks	16.5
Service downlinks	16.5
Feeder uplinks	57.5
Feeder downlinks	57.5
Intersatellite links	Not applicable

Orbits

The Globalstar system uses a (48,8,1) Walker constellation, as detailed in Table 9.5(c). Globalstar integrates deeply into the national or regional PLMN networks; the Globalstar gateways are associated with Globalstar mobile switching centers (MSCs), which are almost identical to and integrated with the existing PLMN MSCs. Globalstar will share home and visitor's location registers (HLR and VLR), numbering, and so on, with the national or regional PLMN, and as such it could be, in principle, completely transparent to the user, whether service is provided by Globalstar or through the PLMN.

9.2.5 Iridium[1] (Iridium Inc.)

Iridium is a global, digital, portable personal communications system in which subscribers use portable or mobile radio units with small antennas to reach a constellation of 66 satellites. The system provides voice, data, positioning, and paging services through the digitally switched constellation. The system provides, generally, line-of-sight coverage to and from every point on the surface of the Earth. The frequency bands and bandwidth required for the Iridium system are shown in Tables 9.6(a) and 9.6(b).

Table 9.5(c)
Globalstar Constellation

Orbit	Altitude (km)	Period (hr)	Inclination (deg)	Number of Planes	Satellites Per Plane
Inclined Circular	~1400	~2	52	8	6

1. Iridium is a trademark and service mark of Motorola Inc.

Table 9.6(a)
Iridium Frequency Bands

	Bands
Service uplinks	1616.0 to 1626.5 MHz
Service downlinks	1616.0 to 1626.5 MHz
Feeder uplinks	Within 27.5 to 30 GHz
Feeder downlinks	Within 18.8 to 20.3 GHz
Intersatellite links	Within 22.55 to 23.55 GHz

Table 9.6(b)
Iridium Bandwidth Requirements

	Bandwidth Requirements (MHz)
Service uplinks	10.5
Service downlinks	10.5
Feeder uplinks	100.0
Feeder downlinks	100.0
Intersatellite links	200.0

9.2.5.1 Coverage and Spot Beam Configurations

Each satellite generates 48 beams to form a continuous overlapping pattern on the Earth; the entire constellation generates about 2,150 active beams. The 48-cell L-band footprint is shown in Figure 9.3.

9.2.5.2 Orbits

The Iridium system uses a six-plane circular polar constellation, as outlined in Table 9.6(c).

9.2.5.3 Air-Interface

The Iridium system provides an L-Band link between the subscriber and the satellite constellation occupying a band of 10.5 MHz. The access format is frequency division multiplex/time division multiple access for data transmissions (FDM/TDMA) and vocoded voice. The width of each frequency channel is 41.67 kHz, which includes 31.5-kHz occupied bandwidth and guard bands. The maximum number of L-Band service duplex channels per satellite is 3,840.

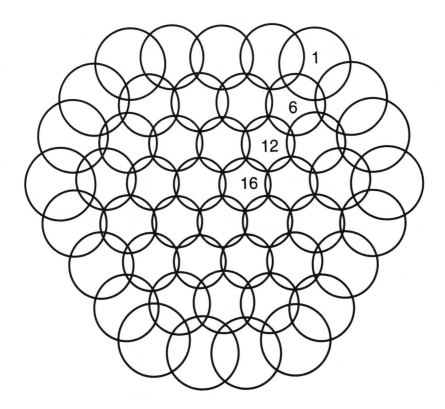

Figure 9.3 IRIDIUM 48-cell L-band footprint.

Table 9.6(c)
Iridium Constellation

Orbit	Altitude (km)	Period (hours)	Inclination	Number of Planes	Plane Spacing (deg.)	Satellites Per Plane
Polar Circular	~780	~1.7	86°	6	31.6 or 22.0	11

The same frequency bands are used for both service uplink and service downlink transmissions. The 90-ms TDMA frame is divided into four transmit and four receive time slots, which are controlled so that the Iridium Subscriber Unit (IRU) burst transmissions arrive at the satellite in proper sequence, with Doppler correction. The downlink format is similar.

The modulation scheme is differentially coded, raised–cosine filtered, quadrature phase shift-keyed (QPSK) modulation, with a burst rate of 50 kbit/s (25 ksymbols/s) i either direction. This format has been chosen as the best compromise for the transmissio channel between satellites and the earth stations, which is subject to a combination o multipath fading and other transmission impairments.

Each subscriber unit operates in burst mode on an assigned frequency channel Since the same frequency is used for the uplink and the downlink, the transmissions ar offset in time. The TDMA frame lasts 90 ms and has four receive and four transmit tim slots, which are synchronized so that each subscriber unit transmission is received at th satellite at the proper time. Two subscribers terminals talking to each other in the sam satellite footprint could use the same frequency channel or two different frequency chan nels. The link is constituted by a single hop established through the satellite, which als controls intersatellite handovers that occur during the call.

9.2.6 Loopus (Germany)

As well as being the name of a special HEO constellation in which handover betweer descending and ascending satellites is performed at practically the same celestial points. LOOPUS is also the name of an S-PCN proposed system utilizing LOOPUS-type HEOs

There are proposals by Deutsche Aerospace/Germany (filing under the name LOO-PUS Mobile D) and also by a LOOPUS initiative in Germany that is not yet institutionalized but is expected to become an international consortium (filing under the name QUASIGEO)

9.2.6.1 Frequency Bands

The frequency bands required for LOOPUS Mobile D and for QUASIGEO-L1, L2, and L3 are given in Tables 9.7(a) and 9.7(b).

Table 9.7(a)
LOOPUS Mobile D Frequency Bands

	Band
Service uplinks	Ku-band under RR 859
Service downlinks	Ku-band on a noninterference basis
Feeder uplinks	Ku-band
Feeder downlinks	Ku-band
Intersatellite links	Not used

Table 9.7(b)
QUASIGEO Frequency Bands

	Band
Service uplinks	Ka-band/L-band/ FPLMTS-band
Service downlinks	Ka-band/L-band/ FPLMTS-band
Feeder uplinks	Ku-band
Feeder downlinks	Ku-band
Intersatellite links	Not used

9.2.6.2 Orbits

The LOOPUS systems will most likely use three-plane HEO constellations in 14.4-hour orbits, providing five quasi positions above the northern hemisphere and repeating their common subsatellite track in 72-hour intervals. Each two neighboring quasi positions are so close that mutual backup is possible in case of failure. The constellation parameters are expected to be as outlined in Table 9.7(c).

9.2.7 Odyssey (TRW)

Odyssey is a satellite systems that will provide high-quality wireless communications services, through terrestrial cellular compatible handsets, to users worldwide, providing them with services that include voice, data, paging, radiodetermination, and messaging.

9.2.7.1 Frequency Bands

The frequency bands and bandwidth required for the Odyssey systems are given in Tables 9.8(a) and 9.8(b).

Table 9.7(c)
LOOPUS Constellation

Orbit	Apogee (km)	Perigee (km)	Period (hr)	Inclination	Number of Planes	Plane Spacing (deg)	Satellites per Plane
LOOPUS HEO	41,449	5784	14.4	63.4	3	120	3

Table 9.8(a)
Odyssey Frequency Bands

	Band
Service uplinks	1610.0 to 1626.5 MHz
Service downlinks	2483.5 to 2500.0 MHz
Feeder uplinks	Ka-band
Feeder downlinks	Ka-band
Intersatellite links	Not used

Table 9.8(b)
Odyssey Bandwidth Requirements

	Bandwidth Requirement (MHz)
Service uplinks	6.5
Service downlinks	16.5
Feeder uplinks	Information not available
Feeder downlinks	Information not available
Intersatellite links	Not used

9.2.7.2 Orbits

The Odyssey system uses a three-plane inclined circular constellation, as outlined in Table 9.8(c).

TRW states that the constellation guarantees a minimum line-of-sight elevation angle of at least 30 degrees to at least one satellite visible in every location for more than 95% of the time.

9.2.8 Inmarsat-P (International Maritime Satellite Organization)

Inmarsat-P is a proposed handheld global satellite phone service, currently under development by the International Maritime Satellite Organization (Inmarsat). Inmarsat-P is still

Table 9.8(c)
Odyssey Constellation

Orbit	Altitude (km)	Period (hours)	Inclination (deg)	Number of Planes	Plane Spacing (deg.)	Satellites per Plane
Inclined Circular	10,354	~6	55	3	120	4

in the midst of a detailed final phase of studies relating to system and market issues, and various concepts and configurations are being analyzed prior to the completion of the service and network definition, leading to Inmarsat-P implementation. Because of this, it is difficult to produce a description of Inmarsat-P in a format entirely analogous to that used to describe the other system proposals. The approach adopted, therefore, is to follow as closely as possible the outline structure used to describe other systems, but to adapt this where necessary to encompass descriptions of the broad ranges of options currently being studied by Inmarsat.

9.2.8.1 Space Segment

Inmarsat are currently studying three orbital configuration options for the implementation of Inmarsat-P: geostationary, intermediate circular, and low earth orbit configurations. A number of the space segment elements are dependent on the orbital configuration.

9.2.8.2 Frequency Bands

The baseline frequency bands and setimated system bandwidth requirements being considered for the Inmarsat-P system are given in Tables 9.9(a) and 9.9(b).

9.2.8.3 Orbits

Inmarsat is investigating a number of different orbital configuration options, as outlined in Table 9.9(c).

Table 9.9(a)
Inmarsat-P Baseline Frequency Bands

	GSO	*ICO*	*LEO*
Service uplinks	1616.5–1626.5 MHz possibly down to 1610.0 MHz alternatively 1980.0-2010.0 MHz		
Service downlinks	2483.5–2500.0 MHz alternatively 2170.0–2200.0 MHz		
Feeder uplinks	6425.0–6575.0 MHz	14.0–14.5 GHz alternatively options at C, Ku and Ka-band	
Feeder downlinks	3600.0–3650.0 MHz	10.95–11.2/11.45–11.7 GHz alternatively options at C, Ku and Ka-band	
Intersatellite links	If ISLs are required they will be in the Ka-band ISL allocations		

Table 9.9(b)
Estimated Inmarsat-P Bandwidth Requirements

	GSO	ICO	LEO
Service uplinks	5 MHz but up to 10 MHz in peak traffic areas		
Service downlinks	5 MHz but up to 10 MHz in peak traffic areas		
Feeder uplinks	Information not provided	200 MHz for CDMA 50 MHz for TDMA	
Feeder downlinks	Information not provided	200 MHz for CDMA 50 MHz for TDMA	
Intersatellite links	Not yet determined		

Table 9.9(c)
Inmarsat-P Orbital Configuration Options

Orbit Option	Altitude (km)	Period (hours)	Inclination (deg)	Number of Satellites
GSO-P circular	~35,786	~24	0	4
GSO-C circular	~35,786	~24	0	4
ICO circular	~10,355 or ~13,892	~6 or ~8	50.7	15 or 12
LEO circular	~1800	~2	55	54
LEO-ISL circular	~1800	~2	90	54

For the ICO constellation option, two or more satellites will always be visible to a user within the service area at an elevation angle of ~20 degrees or greater. The ICO constellation parameters would provide minimum acceptable satellite visibility, but other constellations providing similar or better visibility, with heights in this range, may be considered.

For the LEO constellation option, two or more satellites will always be visible to a user within the service area at an elevation angle of ~10 degrees or greater and one or more at an elevation angle of ~20 degrees or greater. The LEO constellation parameters would provide minimum acceptable satellite visibility, but other constellations providing similar or better visibility may be considered.

9.2.9 MAGSS-14 (European Space Agency)

One of the most recent proposals for a global satellite mobile system for personal communications, the MAGSS-14 system, utilizes an MEO constellation with a total of 14 satellites. It is designed to support handheld terminals with 0.5W transmit power and vehicle units with 7-dB antenna gain and 2W transmit power capability [5].

The MAGSS-14 constellation is similar to that proposed by Odyssey but with a different distribution of orbit planes and satellites in them (Table 9.10).

The constellation is optimized to enhance coverage over Europe, and calculations indicate that between 30 and 60 degrees latitude the elevation angle is above 30, 40, and 50 degrees for 100%, 95%, and 75% of the time, respectively. Nevertheless, elevation angles anywhere on the Earth's surface do not fall below 28.5 degrees at any time during 24 hours.

The system architecture proposes a minimum of 14 gateway Earth stations all connected to a single network control centre.

Uplinks and downlinks to mobiles are envisaged in L-band with Ka-band utilized for links with the Earth ground stations. While from many points of view links in C-band would be preferable, it is doubtful whether satisfactory coordination with GEO C-band users could be achieved.

An initial startup constellation with a single satellite in each of the seven orbital planes appears feasible, since this would still give global coverage at more than 7.5 degrees elevation, and in latitudes between 15 and 65 degrees an elevation angle of more than 30 degrees is achieved for more than 80% of the time.

9.3 INTEGRATION OF SATELLITE AND TERRESTRIAL PCS

The evolution of satellite personal communications has, until relatively recently, been based on the perception that there exists a substantial niche market for standalone services—extensions of the INMARSAT approach. It has become increasingly apparent,

Table 9.10
MAGSS-14 Constellation

Orbit	Altitude (km)	Period (hr)	Inclination	No. of Planes	Satellites per Plane
Inclined Circular	10,350	6	56°	7	2

however, that if satellite PCS is to provide complementary services to the terrestrial equivalent, providing coverage in the gaps between land-based PCS "islands," then interworking between the space segment and terrestrial PLMNs is essential.

There are basically two ways in which such integration can be achieved: to accept that satellite and terrestrial elements will remain separate as far as their air-interfaces and network control are concerned, with interconnection only within the fixed network; or to aim for full integration of space segment with terrestrial PLMNs, sharing the same procedures, protocols, and databases. Trends over the last three to four years have been strongly toward the latter approach [6,7].

Implementing such close integration is, however, not straightforward at either the air-interface or system architecture levels. GSM represents an obvious starting point as far as air-interface considerations are concerned, but it is immediately apparent that utilizing GSM-transmitted signal structure in the space segment would be extremely inefficient. The transmitted bit rate in GSM of 271 kbit/s only supports (8)(13) = 104 kbit/s of useful input digitized speech, a substantial part of the remainder being required to combat Rayleigh fading. In satellite PCS, it is unrealistic to expect service to be maintained when no line of sight from hand terminal to satellite is available, so the GSM physical layer parameters are inappropriate.

This is not a serious difficulty, though, in the context of third-generation PCS since FPLMTS assumes that user terminals will have the capability to adopt whatever combination of air-interface parameters is appropriate to the service and radio environment being accessed at any one time; the satellite link requirements would, therefore, be simply another parameter set among others. System architecture issues present greater difficulties, however, since fundamental choices about the relationship of terrestrial and satellite system entities have to be made. These are likely to significantly affect the ways in which services are licensed and on the techniques deployed to enable users to migrate from terrestrial networks to satellite segments and back again. Two basic possibilities illustrate this point.

In Figure 9.4 one possibility is shown in which the satellite service is supported effectively by a separate PLMN with its own mobile switching centers (MSCs) connected to the satellite hub station. Interconnection with other PLMNs is also achieved at this level. Figure 9.5 shows an alternative arrangement in which the satellite hub station looks like a cell site within a PLMN so that the satellite component of the system shares the MSCs of the terrestrial network and, by implication, shares other entities in the terrestrial PLMN such as home and visitor location registers.

The systems discussed in this chapter lend themselves to the different strategies in varying degrees. Iridium, for instance, envisages the extensive use of intersatellite links with relatively few Earth hub stations— perhaps two or three for Europe—whereas Globalstar proposes a much less ambitious space segment without intersatellite links but employing more numerous hub stations on the ground—two or three per country in Europe. A further factor to take into account is the question of handoff between terrestrial and satellite-supported cells. If it is intended that no handoff is needed, that is, that calls are originated and sustained in either the ground station network or the satellite, then the

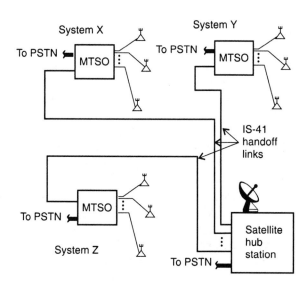

Figure 9.4 Satellite hub station connected to cellular as MTSO.

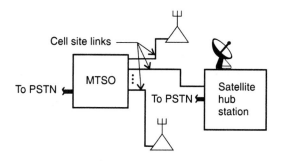

Figure 9.5 Satellite hub station connected to cellular as cell site.

satellite PLMN approach is appropriate. On the other hand, if terrestrial-satellite handoff is to be built into the system, then the second alternative would permit more straightforward handover procedures to be implemented.

These issues are far from being resolved at the time of writing, as are the problems of establishing a licensing regime for the LEO and MEO systems, which are capable of providing global coverage at their inception. Clearly, the constellation owners who put such a system in place will want to receive revenue benefit from users all over the world, but who operates the hub stations and how landing rights to mobiles might be secured remain the subject of negotiations that are as yet in their infancy.

9.4 CONCLUSIONS

In reviewing the various proposals outlined in this chapter, it is evident that there remain different underlying intentions that have motivated the system designers. At the one end of the service spectrum are systems like Iridium that are conceived as standalone platforms for service provision but that could integrate to some extent with terrestrial networks. The next level of system concept, epitomized by Globalstar, aims at a system design that is perceived from the outset as an extension of terrestrial networks and consequently targets a high level of integration. A third approach, which was not featured in the system descriptions here, follows the global service provision capability of large LEO constellations to the ultimate conclusion that such a satellite constellation might provide general telecommunications services to fixed, mobile, and personal terminals in the role of global PTO. Such a system might provide a solution to the otherwise intractable problem of service provision in the developing world where telecommunications infrastructure will take decades to become established by conventional terrestrial network techniques. Even in such environments, however, large user capacity would be a prerequisite, and visibility of the spacecraft would need to be virtually universal in any type of environment, including urban centers. The implications of these requirements are that uplink and downlink operation to all terminals would have to migrate to K/Ka band (20/30 GHz) where sufficient bandwidth is available to support demand and also that a very large constellation of satellites would be needed. Such a system has now been proposed by Calling Corporation [8], which has made public its plans for a constellation of 840 active satellites with 10% standby, giving 924 satellites in total distributed in 40 orbit planes.

While this might appear far-fetched in comparison with the aspirations of, for instance, the Globalstar system, there can now be little doubt that satellite components of future comprehensive personal communications systems will have a significant influence on overall system architecture and will be indispensable to the achievement of FPLMTS objectives.

REFERENCES

[1] Watson, P. A., J. G. Gardiner, A. P. Seneviratne, "Proposal for a Land Mobile Satellite Communications Experiment," *IEE Int. Conf. Mobile Radio Systems and Techniques*, University of York, Sept. 1984, Pub 238, pp. 126–130.
[2] Stuart, J. A., " 'ARCHIMEDES' Mobile Services for Europe From Highly Inclined Orbits: A Cost Effective Solution," *Conf Satellite Communications and Broadcasting 87*, Proc. Online Publications, 1988 pp. 45–57.
[3] Ballard, A. H., "Rosette Constellations of Earth Satellites," *IEEE Trans. on Aerospace and Electronic Systems*, Sept. 1980, pp. 656–665.
[4] Gardiner, J. G., "Regulatory Considerations in Satellite Mobile Systems," *COST 227/COST 231 Joint Workshop*, Limerick, Sept. 1993.
[5] Benedicto, J., J. Fortuny, P. Rastrilla, "MAGSS-14: A Medium Altitude Global Mobile Satellite System for Personal Communications at L-Band," *ESA Journal*, Vol. 16, Jan. 1992, pp. 117–133.

[6] Drucker, E. H., "Integration of Mobile Satellite and Cellular Systems," *Proc. 3rd Int. Mobile Satellite Conf., MSC '93*, Pasadena, CA, June 16–18, 1993, pp. 119–124.

[7] Del Re, E., et al., "Architectures and Protocols for an Integrated Satellite-Terrestrial Mobile System," *Proc. 3rd int. Mobile Satellite Conf., MSC '93*, Pasadena, CA, June 16-18, 1993, pp. 137–142.

[8] Tuck, E. F., et al., "The Calling Network: A Global Telephone Utility," *Space Communications*, Vol. 11, 1993, pp. 141–161.

Part III
The Way Ahead: Evolution and Convergence

Chapter 10

A North American Perspective

Jesse Russell and Anil Kripalani

As part of this chapter on a North American perspective on personal communications, we will describe services envisioned to be part of personal communications services (PCS) and a *zonal* service model to put these in context. Before any discussion on PCS or personal communications networks (PCNs) realization can take place, an understanding of the regulatory environment is necessary. We will present background on the Federal Communications Commission's (FCC) Notice of Proposed Rulemaking (NPRM) that solicited and received comments from a wide variety of industry participants and potential entrants, followed by a comprehensive summary of the FCC's Second Report and Order on spectrum allocations and licensing rules in the emerging technologies band. We will then discuss PCS standardization activities in North America. This information on standards will describe the activities of the Telecommunications Industry Association (TIA) and Committee T1, outlining the structure of these standards organizations and their scope. We will then describe a model for future personal communications networks that draws on the integration of wireline and wireless communications networks. Finally, we will present a brief account of some PCS field trials that have been conducted in the United States.

0.1 THE EMERGENCE OF THE PERSONAL COMMUNICATIONS INDUSTRY

In North America, we have seen rapid and sustained growth over the past five years in most segments of the wireless communications industry, specifically in cordless, cellular, and paging. This growth has given rise to the view that in this large market, we have extended the boundaries and are entering into a new era in communications, characterized by a significant paradigm shift. This change is taking us from the current state of providing *universal telephone service*, which is built on a traditional user location-based communications infrastructure and location-dependent services, to a new anticipated state of supporting *universal personal communications* (UPC) (Figure 10.1), which necessitates the provisioning of a revamped, *person-based* communications infrastructure offering personalized services.

Throughout the world, the definition and concept of personal communications are still being determined by consumers and by the industry. There is general agreement, however, that personal communications will provide the consumer with the ability *to communicate anywhere, anytime, and in any form.* The concept of UPC is built on paradigm shifts in the following key dimensions [1]:

- *User access*, where we are moving toward ubiquitous wireless access;
- *Communications environment*, where the predominant characteristic is mobility;
- *Service flexibility*, where services are geared to be user environment independent;
- *Communications network intelligence*, where the critical intelligence in the network will no longer be in the core of the network but will reside instead in the intelligent servers that hang off the superhighway-like core transport network and in intelligent user terminals. In other words, the network will emulate the client-server model, and intelligence will be focused along the periphery of the network.

Figure 10.2 shows a diagrammatic representation of these dimensions.

Changing lifestyles and demands have placed people on the go in the 1990s. This fact has lent further impetus to the advent of personal communications. Despite extensive research activities, many technological breakthroughs, and proposals for new technologies, market studies in North America indicate that consumers are more concerned with having their needs addressed and not so much with the technology used to address those needs.

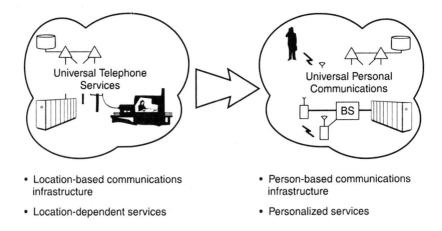

- Location-based communications infrastructure
- Location-dependent services

- Person-based communications infrastructure
- Personalized services

Figure 10.1 A paradigm shift in communications.

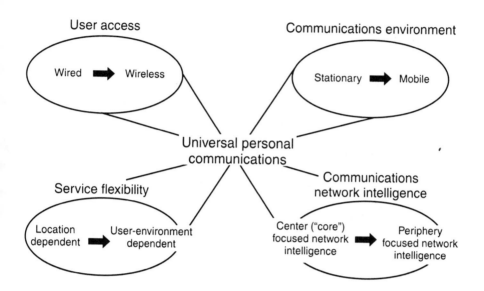

Figure 10.2 The evolution toward universal personal communications.

The vision of UPC will enable users to communicate from anywhere, at anytime, and in any form. This concept refers to basic telecommunication services as well as advanced personalized services through integrated wired and wireless access. Systems that support UPC will be enhanced by the utilization of the integrated services digital network (ISDN) and will be interconnected with other PCNs and non-PCNs, such as the public switched telephone network (PSTN), cellular networks, and premises-based wired networks. Also encompassed under the vision of UPC are low-cost, customer-friendly terminals in the form of pocket-sized radio units with smart-card capabilities, capable of supporting voice, data, imaging, and eventually video services [2].

UPC integrates a variety of telecommunications services that allow people to communicate independent of location, access method, or information format. The concept of *universal service*, first brought up in the context of telecommunications in the United States by Theodore Vail, implies that service providers must provide service access for a broad segment of the population, indeed, to all who may desire such access.

10.2 A PERSPECTIVE ON UNIVERSAL PERSONAL COMMUNICATIONS

In this section, we will describe the benefit customers must be able to derive from UPC, how the services might evolve, and the extent to which those services will resemble existing wireless services.

10.2.1 The Consumer Value Proposition

A popular debate in the North American wireless industry, like elsewhere across the world, has focused on whether PCS will originate from or be derived from cellular-mobile technology and services or, alternatively, from cordless technology and services. The two "origins," that is, cellular and cordless, differ most significantly in the degree of mobility and the price of service. In response to these claims, we submit that PCS will, in fact, draw on the characteristics and experiences of both cellular technology and cordless technology. In addition, we see that PCS will draw on paging, satellite and specialized mobile radio (SMR) technology as appropriate (Figure 10.3). We also believe that PCS will draw on intelligent network (IN) technology and services for mobility management capabilities. The important point to note is that the vision of UPC will be derived from this blending to offer the consumer the following benefits:

- Universal services;
- Transparent mobility;

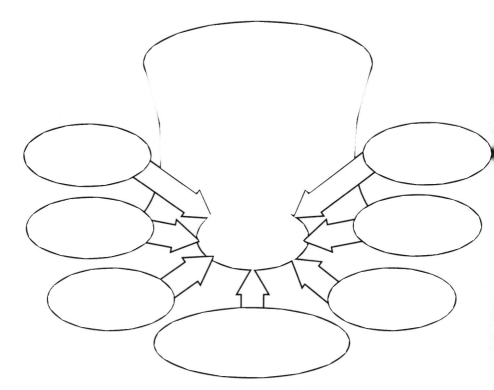

Figure 10.3 The consumer value proposition in communications.

- Low cost;
- High quality;
- User-friendly interfaces.

Although this chapter will explore the architectural aspects of UPC, emphasis should remain on the service aspects that the subscriber sees, keeping in mind that the subscriber has played, and will continue to play, a significant role in creating and supporting this market.

10.2.2 Personal Communications Services

It is expected that our vision of UPC will apply to high-mobility, low-mobility, and fixed wireless environments. UPC should eventually support broad-bandwidth services, such as high-quality voice, high-speed data, high-resolution imaging, and full-motion video communications. However, one might expect there to be tradeoffs between mobility required and bandwidth supported. Services in the low-mobility or fixed wireless in-building environments will include the full range of high-quality voice, high-speed data, imaging, and eventually multimedia capabilities, taking advantage of low-power picocellular configurations. The low-mobility outdoor or pedestrian environment would utilize a low-power microcellular network infrastructure and support a set of services that include high-quality voice, medium-speed data, and two-way messaging. The range of services in high-mobility, vehicular environments will be supported by systems that rely on larger macrocells with higher power base stations that handle high-speed handoffs. Services in such environments will include good-quality voice services, low-speed data services, and two-way messaging.

Personal communications services are expected to require convenient and user-friendly pocket-sized devices, which will enable users to communicate or be reached without any constraints for a wide range of voice, data, text, and visual services across multiple environments. The network will need to distinguish the type of environment in which the subscriber is operating and provide services accordingly.

Potential Market Segments and Applications

There are several major market segments where the availability and flexibility of wireless technology permit the advent of new and, in many cases, nontraditional solutions and service offerings. These key market segments and some of the applications worth considering under each segment are listed in Table 10.1.

10.2.3 Proposed User Service Environment Model

To facilitate the advent of UPC, a multienvironment service model is helpful to characterize communications in a variety of different environments. Such a model must consider many

Table 10.1
Market segments and corresponding applications

Key Market Segment	Applications
Communications networks	Private corporate wireless networks Wireless Centrex Wireless LAN/WAN
Retail	Personal financial card
Financial	Wireless point-of-sale
Transportation	Automatic toll collection Personal navigation Traffic advisories
Entertainment	On-site event Interactive network gaming
Education	Wireless library services Mobile scholar
Healthcare	Emergency medical services (EMS)
Mobile information services	Field operations Repair and delivery services

factors: maximum distance from a radio distribution point or coverage area, expected traffic density, speed of mobility, and service feature requirements (e.g., bandwidth). Such a service model would support the user's need to communicate by enabling the following capabilities:

- *Universal information access*, which takes advantage of a variety of wireline and wireless access media and permits users to get service wherever they are;
- *Transparent mobility management*, which supports terminal mobility, personal mobility, and service mobility;
- A *personalized service environment*, which supports user control over available services, as well as customization of the way services are presented, while allowing the ability to have access to and modify the user profile.

In the rest of this section, we describe such a multienvironment model for UPC that describes six service zones of application. The six zones are illustrated in Figure 10.4. The term *zone* reflects the type of service environment and span of coverage.

Users would subscribe to grades of service based on the service features they need, the access terminals they want to use, and, most significantly, the environments in which they want to use the service. For the individual user, service across different environments represents the full view of personal communications. Let us briefly review these suggested service zones.

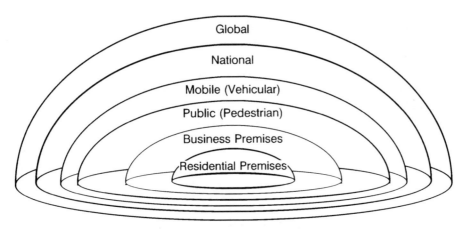

Figure 10.4 A zonal service environment model for universal personal communications.

- *The Residential Service Environment (Zone 1).* This environment will support a variety of voice, data, imaging, and multimedia services in a residential or in-building context. The primary source of radio frequency (RF) distribution will be a radio port–personal base station, owned by the residents and identified with a unique signature. Multiple channels could be available if necessary in this configuration. The full complement of services will be limited to the coverage range of this type of radio port–base station.
- *The Business Service Environment (Zone 2).* This environment pertains to business-oriented, in-building, campus service and is also expected to support a full complement of high-quality voice, high-speed data, high-resolution imaging, multimedia, and video conferencing services through wireless private branch exchange (PBX)/ Centrex configurations and a network of low-range radio ports or picocells.
- *The Public (Pedestrian) Service Environment (Zone 3).* This environment is characterized by low-mobility, urban or suburban, neighborhood-based wireless communications. This zone may also support high-quality voice services, medium-speed data, imaging, fax, and text. Microcellular networks would typically be expected to support coverage in such a zone.
- *The Mobile (Vehicular) Service Environment (Zone 4).* Services in the mobile zone would compare with those provided by cellular today and be characterized as high mobility in urban, suburban, and rural contexts supported by macrocellular networks. The typical user terminal would be capable of medium-quality digital voice and low-speed data. Services for intelligent vehicle highway systems (IVHS) [2], facsimile, and other low-speed data applications would be supported for end users.
- *The National Service Environment (Zone 5).* This zone may be served by megacells (coverage footprint of the order of 100 to 150 miles), which may require the use of

satellite access (e.g., low earth-orbiting systems). Coverage with such megacellular networks would be nationwide. Near-term services provided would be voice and low-speed data.

- *The Global Service Environment (Zone 6).* The global zone type service will support countrywide or transcontinental wireless access. Providing coverage may require a satellite infrastructure, specifically since users may be in areas where terrestrial systems do not exist or cannot support service access. Typical users would be airline, marine, and railroad passengers, as well as motorists who require service in remote areas. This zone will be served by megacells, providing global coverage. Near-term services will consist of voice and low-speed data. It must be stressed that the services in all these zones fall under the overall umbrella of UPC.

10.3 THE U.S. PCS REGULATORY ENVIRONMENT

10.3.1 Background

In late 1989, Petitions for Rule Making[1,2] were filed with the FCC, the regulatory agency responsible for U.S. frequency allocations used by the public. These filings began a series of regulatory and legislative actions and broad industry activities related to the development of PCS in the 1.8 to 2.2 GHz band[3] of the spectrum. The FCC responded quickly to these early filings and in 1991 issued a Notice of Inquiry[4] (NOI), which recognized the public benefit of developing PCS and that the establishment of frequency allocations for PCS was warranted. For providers to be able to quickly and economically deploy services in existing and new markets, the Commission emphasized the importance of providing spectrum and a regulatory framework. In addition, the objectives of *universality, speed of deployment, diversity of services, and competitive delivery* were outlined, and the Commission stated its intent to balance them in providing the regulatory structure and spectrum for PCS.

To establish PCS at 2 GHz, regulatory issues that needed to be addressed included the following:

- PCS definition (from the perspective of the FCC);
- Spectrum allocations at 2 GHz;
- Amount of spectrum for *licensed* operations (requiring explicit authorization to radiate) and *unlicensed* operations (where users can radiate without a need for explicit permission);
- Number of service providers to be licensed per service area;

---

1. Federal Communications Commission, RM-7140, September 1989.
2. Federal Communications Commission, RM-7175, November 1989.
3. This band is frequently referred to as the 2-GHz band. This terminology will also be used in this chapter.
4. Notice of enquiry, GEN Docket No. 90-314, 5 FCC Rcd 3995 (1990).

- Size of licensed service areas;
- Eligibility requirements for licensees;
- Licensing mechanisms;
- Regulatory classification of PCS providers;
- Technical standards;
- Displacement of and impact on incumbents in the affected frequency bands.

ubsequent to the NOI, to which more than 150 parties submitted comments or reply omments, the FCC issued a Policy Statement and Order,[5] a Notice of Proposed Rule laking and Tentative Decision,[6] and a Tentative Decision and Memorandum Opinion nd Order[7] and held an en banc hearing.

Over 220 PCS experimental licenses were authorized by the FCC during the four-ear period from 1989 to 1993. Trials included those for development and testing of quipment across a wide range of spectrum as well as market studies related to a variety f technologies and service concepts.

Legislative actions were also quickly brought to bear on PCS. The Omnibus Budget teconciliation Act of 1993 addressed the method by which PCS spectrum could be warded and the regulatory treatment of personal communications services. The Omnibus udget Reconciliation Act also assisted the FCC in its speed of deployment objective by uthorizing the Commission to employ competitive bidding procedures to choose licensees o use the PCS spectrum. Congress also directed the FCC to promote the objectives set orth in the Communications Act,[8] namely, to benefit all the people of the United States nd, significantly, directed the Commission to provide safeguards that "will promote conomic opportunity and competition and ensure that new and innovative technologies re readily accessible to the American people by avoiding excessive concentrations of icenses and by disseminating licenses among a wide variety of applicants, including small usinesses, rural telephone companies, and businesses owned by members of minority roups and women."[9]

0.3.2 The FCC's Second Report and Order

)n September 23, 1993, the FCC issued its Second Report and Order,[10] in which spectrum vas allocated for PCS[11] and licensing rules and issues were described. The Commission

---

. Policy Statement and Order, GEN Docket No. 90-314, 6 FCC Rcd 6601 (1991).

. Notice of Proposed Rule Making and Tentative Decision, GEN Docket No. 90-314 and ET Docket No 92-100, 7 FCC Rcd 5676 (1992).

. Tentative Decision and Memeorandum Opinion and Order, GEN Docket No 90-314, 7 FCC Rcd 7794.

. U.S.C., Section 151.

. 1993 Budget Reconciliation Act, Section 6002(a).

0. Second Report and Order, GEN Docket No. 90-314 FCC 93-451 (1993).

1. In addition, the allocation for Emerging Technologies was specified under Docket 92-9.

provided a broad definition of PCS in the report, concluding that an ''open, flexible' definition would provide an effective means of meeting the four objectives of *universality speed of deployment, diversity of services, and competitive delivery.* In the following quote, the Commission defined PCS as a family of services not bounded by any particular technology, such as broadband or narrowband access:

> Radio communications that encompass mobile and ancillary fixed communication services that provide services to individuals and businesses and can be integrated with a variety of competing networks.[10]

The FCC considered many factors in reaching their conclusions regarding licensing issues, and stated[10] that a number of factors had to be balanced, including:

- The need to provide sufficient spectrum per license so that viable services could be developed;
- Sharing of spectrum with incumbent fixed microwave operations for some licensees;
- Sufficient number of licenses to ensure a competitive environment;
- The need for spectrum for unlicensed personal communications services;
- The need to conserve the limited amount of spectrum available for new uses.

The Second Report and Order ruled on several of the most important regulatory issues related to PCS:

- Spectrum allocations for licensed and unlicensed PCS;
- Geographic coverage of each license (service areas);
- Amount of bandwidth per license;
- Eligibility requirements for each license;
- Licensing mechanism;
- Technical standards, which include the questions of base station and mobile radiated power.

10.3.2.1 Spectrum Allocations

In its September 23, 1993, ruling, the FCC allocated PCS spectrum in the 1850 to 2200 MHz band of the spectrum. A total of 120 MHz was allocated for licensed operation, a bandwidth that is more than twice the amount now provided for cellular service. A total of 40 MHz was allocated for unlicensed operation (Figure 10.5). These allocations and the specifics that fall under them are subject to change under the reconsideration phase that is underway as of late March 1994.

Also, in a separate ruling, spectrum was allocated for narrowband PCS in the 901 to 902 MHz, 931 to 932 MHz, and 941 to 942 MHz bands. The 120-MHz block for licensed operation was allocated as two blocks of 30 MHz, one block of 20 MHz, and four blocks of 10 MHz. This range of bandwidths reflects the Commission's conclusion that 10 MHz, in conjunction with advanced digital modulation techniques and microcellular

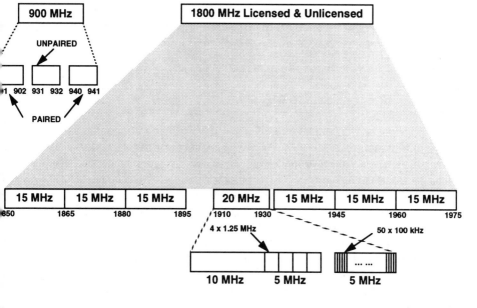

Figure 10.5 FCC spectrum allocations for PCS in the 2-GHz band.

technology, is enough bandwidth for many forms of PCS. The 30-MHz allocations, on the other hand, reflect the expectation that some types of PCS will require more bandwidth. Additionally, the actual frequency bands reflect the relative ease with which the spectrum can be shared with incumbent fixed microwave users.[12]

The Commission stipulated that a licensee cannot aggregate more than 40 MHz in any geographic area. It further stipulated that in determining aggregation limits, ownership interests of 5% or more will be attributed to the holder. Thus, for example, if an entity had an ownership interest in 5% or more of a 10-MHz block, then the entire 10 MHz would be attributed to that entity, and it would be eligible for up to 30 MHz more in that service area. The FCC included aggregation limits so that no single interest can exert undue influence through partial ownership in multiple PCS licenses in a single service area. The 40-MHz limit also ensures that there will be at least three service providers per area and seems to strike a balance between market opportunity and market dominance.

0.3.2.2 Service Area Definition

Prior to the September 1993 Second Report and Order, the FCC had requested comment on four options for service areas to be licensed for PCS. The Commission also suggested

---

12. In the 30-MHz block, fixed microwave uses larger 5-MHz and 10-MHz channels, so a wider bandwidth is needed to effect sharing; in the 10-MHz and 20-MHz blocks, fixed microwave uses smaller 0.8-MHz and 1.6-MHz channels, so smaller allocations will suffice and still allow sharing.

additional options, including determining service areas by competitive bidding and service areas sized differently according to the spectrum blocks licensed. The four options for service area licensing proposed by the Commission were:

1. 194 Local access and transport areas (LATAs);
2. 51 Major trading areas (MTAs);[13]
3. 492 Basic trading areas (BTAs);[13]
4. Nationwide.

Although the FCC did not propose the use of cellular Metropolitan Statistical Area (MSAs)/Rural Statistical Areas (RSAs), some commenting parties expressed a preference for licensing areas that coincided with those used for cellular service. The National Technical Information Agency (NTIA) proposed the use of 183 economic areas defined by the Commerce Department's Bureau of Economic Analysis. These areas generally consist of an MSA and surrounding areas that reflect commuting patterns. The Commission concluded, however, that the use of cellular MSA/RSAs might result in "fragmentation of natural markets,"[10] pointing out that MTAs and BTAs are based on the natural flow of commerce. The MTA/BTA boundaries are drawn on a county-line basis and were determined by detailed studies that considered "physiography, population distribution, newspaper circulation, economic activities, highway facilities, railroad services, suburban transportation," and other factors.[13]

The spectrum allocations are split evenly between BTAs and MTAs, with 60 MHz each. Both 30-MHz blocks are designated for MTAs; the 20-MHz block and the four 10-MHz blocks are designated for BTAs.

10.3.2.3 Unlicensed PCS

The FCC concluded early in the regulatory process[4] that unlicensed PCS was in the public interest and proposed that 20 MHz of spectrum be set aside for its use and that it be shared on a co-primary basis with fixed microwave operations. Many parties argued that 20 MHz was not enough, and several commercial interests and IEEE Project 602 suggested that 35 to 70 MHz was needed.[10]

Comments on unlicensed PCS from the fixed microwave community indicated concerns about unacceptable interference, especially from movable devices. Conversely, companies interested in developing these devices argued that sharing was not feasible because their *nomadic* nature would make mobile devices susceptible to interference by fixed microwave services.

In the Second Report and Order the Commission allocated 40 MHz of spectrum in the 1890 to 1930 MHz band to unlicensed operation. They also required that only digital modulation techniques be employed in this band, limited certain subbands to isochronous-only or asynchronous-only communications, and specified channelization and power

---

13. 1992 Commercial Atlas and Marketing Guide, 123rd Ed., Rand McNally.

requirements. In addition, the FCC required that devices and systems be coordinated through the Unlicensed PCS Ad Hoc Committee for 2 GHz Microwave Transition and Management (UTAM) before being deployed or relocated. Applicants for equipment authorization must be participants in UTAM. UTAM was also designated as coordinator for the transition of this band from fixed microwave service to unlicensed PCS, pending acceptance of an appropriate funding and band-clearing plan. More details are provided in Docket 90-314.[10]

10.3.2.4 Licensing Issues

One of the most significant issues regarding PCS licenses was the question of whether cellular providers and local exchange carriers (LECs) would be allowed or excluded from obtaining licenses. The FCC answered this question in the Second Report and Order by concluding that the public interest would be served by allowing cellular providers and LECs to obtain PCS licenses, subject to specific eligibility requirements that are intended to ensure competition.

Cellular providers are eligible to obtain licenses outside their cellular service areas. They may also obtain licenses within their cellular service area, limited to only one of the 10-MHz BTA frequency block licenses. This limit applies only if there is a 10% or greater population overlap with the cellular service area and the BTA/MTA PCS service area. The FCC placed this restriction on cellular eligibility to guard against the potential for unfair competition if cellular operators provide personal communications services in the same areas where they provide cellular service. The Commission also defined ownership attribution limits so that any party with a 20% or more interest in a cellular entity may not have an attributable interest in a PCS licensee within its area beyond the single 10-MHz BTA license. The Commission appeared to be ensuring that competition between cellular and PCS is not undermined and placed parties on notice that ownership rules would be reconsidered if they were deemed inadequate.[4] The FCC also noted that services like in-building wireless voice services, wireless data transmission services, and telepoint services are included under the PCS definition, and in this connection, the FCC revised cellular rules to explicitly permit cellular licensees to provide any PCS-type services without prior notification.

With regard to LEC entry into offering personal communications services, the Commission created no specific set-aside for LECs. Instead, it made them generally eligible for PCS licenses except in areas where they have attributable cellular interests. In reaching this decision, they concluded that there were compelling reasons to permit LECs to provide PCS and use their existing infrastructure for backhaul functions to benefit the evolution of all personal communications services.

Regarding the licensing process, comparative hearings, lotteries, and competitive bidding were options that were examined. The FCC concluded[10] that the use of competitive bidding to award licenses was the fastest, most efficient, and most economic approach for both government and applicants. Congress had already authorized the Commission to use competitive bidding,[9] and they ruled that that was the mechanism to be employed. The term of PCS licenses will be 10 years, with a high renewal expectancy, similar to that of cellular licenses. The Commission also set build-out requirements, commencing at the time of licensing, to offer service to:

- One-third of the population of the service area within 5 years;
- Two-thirds of the population within 7 years;
- Ninety percent of the population within 10 years.

Failure to meet the build-out requirements will result in forfeiture of the license and permanent ineligibility to regain it.

10.3.2.5 Technical Standards Issues

Convinced that maximum flexibility in technical standards is needed to allow the development of new services, the FCC did not rule on technical standards dealing with air-interfaces, system architectures, or other implementation-specific characteristics. The Commission did, however, set standards for antenna height, coordination distances, and interference-level calculations between PCS-microwave and PCS-PCS. For unlicensed PCS in the 1890 to 1930 MHz band, the 1890 to 1900 and 1920 to 1930 MHz spectrum are limited to isochronous (e.g., voice) operations and the 1900 to 1920 MHz band is limited to asynchronous (e.g., data) operations. A modified version of the Spectral Etiquette proposed by the WINForum organization was adopted by including it in the FCC Part 15 technical standards for PCS. As stated, compliance with the etiquette was to be ensured by requiring FCC certification for equipment. At this time, it is possible that some modifications to these requirements will be made. Radio frequency hazard requirements were also addressed, and the Commission ruled that compliance, with some modifications, to the new Institute of Electrical and Electronics Engineers/American National Standards Institute (IEEE/ANSI) guidelines[14] is required. Base station and mobile power level issues were also discussed. The Commission also urged that participants carefully consider E-911 Emergency Service capabilities in the deployment of personal communications systems and services, pointing out that, for the health and safety of citizens, E-911 capability was of serious concern. The displacement of incumbent users of the 1850 to 2200 MHz band remains a significant issue.

10.3.2.6 Next Steps: PCS Auctions

Although the FCC has ruled on the most significant issues relating to the regulatory aspects of PCS, several important issues and events remain. The 1993 Budget Reconciliation Act

---

14. "Safety Levels With Respect to Human Exposure to Radio Frequency Electromagnetic Fields, 3 kHz to 300 GHz," IEEE C95.1-1991, April 1992. Also adopted by ANSI: ANSI/IEEE C95.1-1992.

directed the FCC to issue a Final Report and Order on Docket 90-314 by February 6, 1994. The Commission met that statutory deadline by adopting the spectrum allocation and operating and licensing regulations for PCS in the Second Report and Order.[10] The Commission was also directed by the Budget Act to commence issuing PCS licenses by May 7, 1994. Many of the broad regulatory hurdles of PCS that fall under the jurisdiction of the FCC seemed to have, to a large extent, been cleared. However, the process is currently (as of March 1994) in the "recon" (reconsideration) phase, and all the specifics of the ruling described are subject to significant change. Our attempt has been to bring the reader up-to-date on the events that have transpired in this arena.

To summarize, it appears that although regulatory details remain, the next year will determine the shape and features of the overall PCS opportunity in the United States.

10.4 PCS STANDARDS ACTIVITIES IN THE UNITED STATES

It has become quite clear that service and equipment standards need to be established for the U.S. personal communications industry. Since personal communications truly has the potential to bring the existing wireline and wireless telecommunications industries together, it becomes important that the various organizations authorized to set standards for voice, data, text, imaging, and visual services and equipment be coordinated in their efforts. If such care is not taken, users may be subjected to multiple and incompatible standards that would lead to the failure of PCS technology in the market.

U.S. standards efforts related to PCS are focused in two major areas: wireless access with terminal mobility and personal mobility. These align with European and international standards activities known as future public land mobile telecommunications systems (FPLMTS) and Universal Personal Telecommunication (UPT), respectively.

The United States has three existing organizations that are accredited by ANSI to establish telecommunications standards and are associated with PCS. We will briefly describe two of these: the Telecommunications Industry Association (TIA), which is affiliated with the Electronic Industries Association (EIA), and Committee T1 because they have a significant initiative in the area of PCS. The third is the IEEE, which is not explicitly targeting PCS standards. However, the IEEE 802.11 subcommittee has done extensive work to develop specifications for high-speed wireless data networking. The IEEE structure and work are not described here.

The Personal Communications Industry Association (PCIA), formerly known as Telocator, is an organization with participation from a broad spectrum of wireless product vendor companies and service providers. They have been active in developing requirements for PCS standards and have produced several reports for use by the wireless industry and regulators.[15,16]

---

5. "Network Interface Standards Requirements Document for Personal Communications, Version 1," Telocator T&E Committee, January 6, 1993.

6. "Data Standards Requirements Document for Personal Communications," Telocator T&E Data Subcommittee, July 1993.

TIA has been in existence for over 50 years and has historically been the standards body responsible for mobile communications standards, fiber-optic equipment standards, and PBX equipment standards, to mention a few. On the mobile communications front, TIA has developed and published analog and digital cellular standards, private land mobile standards, consumer radio and cordless equipment specifications and is now developing 2-GHz PCS standards.

Since its inception in 1984, Committee T1 has focused on wireline access and network standards such as ISDN, signaling system number 7 (SS7), and IN. It is structured functionally, with technical subcommittees (TSCs) on signaling, OA&M, performance, and services, as well as PCS program management. Since PCS is of interest to both T1 and TIA standards organizations, it is being worked in both parallel and joint arrangements.

The structure of the PCS standards committees in TIA and T1 is described below.

10.4.1 PCS Standardization in TIA

As far as the TIA is concerned, there are two sections (800 MHz Section and 1800 MHz Section) and associated technical engineering committees (TR45 and TR46) that have responsibility over PCS standards. The sections are policy bodies that provide guidance to their respective TR Committees.

The Cellular Telecommunications Industry Association (CTIA) has traditionally provided user performance requirements and general guidance to the 800 MHz Section and TR45. TR45 has been in place since 1982 and has been responsible for analog and digital cellular standards. Most significantly, TR45 and its subcommittees have developed the North American Digital Cellular time-division multiple access (TDMA) (IS-54) and code-division multiple access (CDMA) (IS-95) standards for 800-MHz operation. They have also been responsible for the intersystem standards (IS-41) that support intersystem roaming and automatic registration capabilities. The current focus in this committee is mobile and personal communications services at 800 MHz. The following is the list of TR45 subcommittees and their primary function:

- TR45.1, analog cellular standards;
- TR45.2, intersystem operations;
- TR45.3, digital cellular (TDMA);
- TR45.4, microsystems/PCS (800 MHz);
- TR45.5, spread-spectrum digital cellular (CDMA).

TIA's TR46 committee, which began work in February 1993, came into existence as the engineering committee responsible for developing standards for mobile and personal communications services in the 1850 to 2200 MHz band. TR46 has received requirements for standardization from the PCIA, as well as from the CTIA. TR46's current organizational structure consists of three standing subcommittees and an ad hoc subcommittee (Figure 10.6). This structure may be revised at any time in keeping with the needs of standardization in this band. These subcommittees and their responsibilities are as follows:

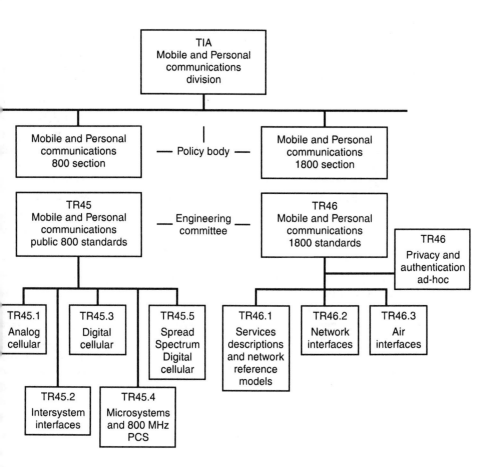

Figure 10.6 TIA mobile and personal communications standards organization (TR45 and TR46 only).

- TR46.1, Services and network reference models;
- TR46.2, Network interfaces - radio subsystem/switch interface 'A', PCN-PCN intersystem operations, PCN/other network entity internetwork operations;
- TR46.3, Air interfaces; responsible for PCS air-interface standards and minimum equipment specifications, also parent organization of the Joint Technical Committee (JTC) with T1P1.4;
- TR46, Ad Hoc on privacy and authentication; establishes algorithms and protocols for the implementation of PCS P&A requirements.

10.4.2 PCS Standardization in Committee T1

Committee T1 was created after the 1984 AT&T divestiture and is a standards body where the interest groups are local exchange carriers (LECs), interexchange carriers (IXCs), equipment manufacturers, and other general-nterest participants. T1 has several TSCs that focus on PCS program management (T1P1), signaling (T1S1), OAM&P (T1M1), performance (T1A1), interfaces (T1E1), and digital hierarchy (T1X1). Figure 10.7 shows the T1 structure.

The T1P1 TSC was formed in late 1990 and chartered to provide program management for complex standardization projects such as PCS. T1P1 established and has maintained a program management team that consists of representatives from various T1 TSCs and that interfaces to TIA TR45 and TR46 as well as PCIA. This cross-organizational group coordinates PCS standards activities across T1 and TIA to eliminate duplication of effort. Working groups T1P1.2 and T1P1.3 are defining high-level requirements for PCS standardization and are documenting these in technical reports, several of which are currently being balloted for approval. These technical reports cover wireless access systems and service objectives and network capabilities, architectures, and interfaces. The reports are intended to be used within T1 and TIA as the basis for developing detailed technical standards.

T1P1 also has played a significant role in defining personal mobility, also known as UPT, which introduces personal numbers as a means by which users can make or receive calls from any type of terminal, wireline or wireless, and be able to manage incoming calls. Internationally, the International Telecommunications Union (ITU, formerly CCITT) is leading the standardization of UPT; ETSI, the European standards organization has parallel activity. The current U.S. approach is to provide input to the ITU and then to adopt the results from the ITU in conjunction with any "deltas" applicable to the U.S. environment.

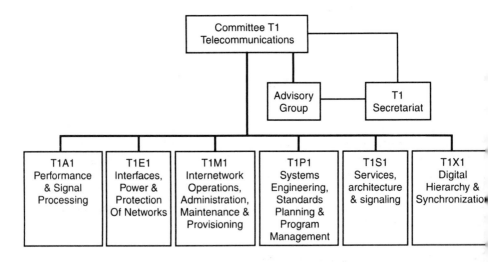

Figure 10.7 Committee T1 personal communications standards organization.

Aside from T1P1, T1A1 has also been active in the area of PCS performance and as already balloted a technical report on this topic. T1M1 is focusing on the area of perations, administration, maintenance, and provisioning for PCS. T1S1 is expected to egin work on standards related to access signaling and networking signaling.

).4.3 The Joint Technical Committee on Wireless Access

Joint Technical Committee (JTC) was formed in 1992 between TIA TR45.4 and T1E1 focus on standardizing the physical layer of the public air-interface for PCS (in the ;50 to 2200 MHz range). The formation of the JTC was preceded by a Joint Experts ·.eeting held in November 1992, the results of which provided a starting point for the ITC's work. After the creation of TR46 and reorganization in T1, the parent organizations)f the JTC are now TR46.3 and T1P1.4. The JTC has defined methodology and selection :riteria to be used in selecting appropriate radio interface technology and is currently :valuating candidate air-nterface proposals. Seventeen candidates were proposed in November 1993 from individual companies and consortia from across the globe. After :onsolidation, seven remain at the current time (March 1994), and the stated intention of 4he JTC is to begin developing the specifications for each of these. It is hoped and expected 4hat the service providers for PCS in the 2-GHz band, once licensed, will drive the market :oward a *small* number of air interfaces, so that the industry can focus on offering cost-:ffective systems and services for the consumer.

10.5 REALIZATION OF THE PERSONAL COMMUNICATIONS NETWORK

The fundamental basis of future PCNs that will support universal personal communications is the provisioning of ubiquitous untethered access, as well as broadband switching and transport, coupled with personalized service management. In this connection, PCNs will depend on the deployment of intelligent radio base stations and microcellular radio port infrastructure for wireless access over different ranges to support a variety of end-user services. Broadband switching and transmission technologies that are now evolving will also play a key role. Developments in intersystem networking, INs, and client-server technology will be instrumental in realizing seamless operation and service management.

A logical representation, or high-level architecture, of a generic PCN to support UPC is shown as a layered model in Figure 10.8 and is described in this section.

This network architecture supports seamless access to personal communications services. The user perceives the network to support ubiquitous access and transparent ·~obility. The realization of the services envisioned within the scope of universal personal)mmunications requires significant capabilities and intelligence at different points in the etwork. As shown in this logical representation, the future PCN is represented in terms f the following layers:

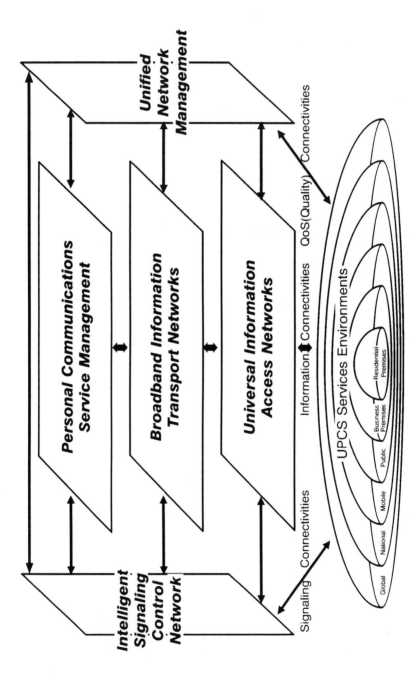

Figure 10.8 A logical view of the future personal communications network.

- Universal information access;
- Broadband information transport;
- Personal communications service management;
- Intelligent signaling control;
- Unified network management.

The *universal information access network layer* provides the functions associated with the wireless access part of the network.

Broadly, this includes those functions necessary to establish and maintain communication between a wireless terminal that may or may not be on the move and the public network, while tracking its location. Specifically, the following functionality is included:

- Air-interface channel coding and decoding and error detection and correction;
- Modulation/demodulation;
- Synchronization and signal amplification;
- Channel processing;
- Signaling and bandwidth management;
- Radio resource management;
- Terminal mobility management (incoming call routing and delivery, handoffs);
- Security, privacy, and authentication.

The *broadband information transport network layer*, in our model, is capable of supporting a full range of potential voice, data, image, visual, and other broadband types of services and applications. In the UPC environment, efficient call processing will be required due to higher user mobility and user density. The requirements of mobility and user location tracking and registration will generate high volumes of non-call-related signaling traffic that will present a substantial load on the underlying core network. The transport network layer will need to meet these needs by incorporating higher bandwidth switching and transport systems and services such as broadband ISDN, frame relay, and asynchronous transfer mode (ATM).

The *personal communications service management layer* provides the management of network services and includes administration and control of interfaces for service deployment, configuration, monitoring, and customization. This layer supports the customization and rapid introduction of new or enhanced services to meet specific customer needs and provides service-modeling capabilities.

The *intelligent signaling control network layer* is expected to support the following functions: virtual circuit/path/channel connectivity between the end user and the network, synchronization, intelligent routing (e.g., shortest path, grooming), and special network service functions (e.g., aid in mobility, paging).

The *intelligent network management layer* is essential to achieve quality of service and optimal monitoring and use of wireless resources. This network layer will provide the full range of OA&M functionality, which includes a centralized, integrated, intelligent control of the entire network.

10.6 A SELECTION OF NORTH AMERICAN PCS/PCN SERVICE TRIALS

PCS trials in the United States have been largely focused on propagation characteristics, spectrum sharing, and technology comparison (frequency-division multiple access [FDMA], TDMA, and CDMA). Although technical limits are the major focus, an understanding of the consumer pulse is also being gained. Several critical customer-driven market trials have focused on the willingness of the consumer to pay for personal communications services, with specific attention to feature and function preferences. This section briefly describes a few of these trials conducted in different parts of the United States. Inclusion of any equipment manufacturers' names by no means implies an endorsement of such selections. Information on service costs has been omitted explicitly, since such data is changeable without notice.

10.6.1 GTE Trial

GTE conducted one of the largest PCN tests in the United States when they committed to provide Tele-Go, an advanced cordless service, to 3,000 users in Tampa, Florida [3–5]. This service could be used in or out of the home or office. The service concept allowed a user to have a designated home territory for billing purposes. Users paid the same rate for service in their home territory and a premium rate to use the phone in other areas. Home base stations permitted subscribers to have two-way service. A number of the subscribers had no basic monthly phone charges because they would be disconnected from their landline service. Some of the users were not able to make calls in the premium service coverage areas and were restricted to their home territory. Users participating in this trial were required to pay a monthly charge for use in the home territory and a per-minute charge for premium coverage.

In Nashville, Tennessee, and Durham, North Carolina, GTE expanded its PCS trial to include wireless wide-area Centrex and PBX services. Five businesses participated in the Tele-Go Business Service using 350 handsets in a 3,500-mi^2 area. The focus was on customer needs first and technology second. Customers were to keep their wired Centrex or PBX phone and telephone number and the wireless handset that serves as a wireless extension. Users were charged a flat monthly rate for service in and around their office complex. When they were away from the office area, a per-minute fee was charged for service provided by GTE's cellular system. GTE expected to learn how wireless communications is used in business environments and hoped to better understand the mobility requirements and features for business customers. Trial results have been presented and GTE has announced the launch of a commercial service [6].

10.6.2 Ameritech Trial

This Ameritech experiment in Chicago involved 1,000 randomly selected users [7]. Users lived, worked, and shopped in the covered areas. The trial was a technology trial and a service trial. Initially, the trial employed CT2 equipment using FDMA–time division duplex (TDD) technology. Later, modified digital European cordless telecommunications (DECT) equipment was expected to be used. Services incorporated into the trial included outgoing calling, paging, two-way calling, and voice mail. Users could make and receive calls on a 6.5-oz portable phone with an integrated pager. Base stations supported FDMA or TDMA techniques and connected to the network through an ISDN basic rate interface. The trial was intended to make use of the distributed intelligence capabilities of Ameritech's IN for user registration, user location, call routing, and delivery. Handoff was performed by a central office equipped with intelligent network hardware and software. A service control point (SCP) equipped with IN and additional software was to be used to support call routing, user locating, and users PCS service profiles.

10.6.3 Bell Atlantic Trial

Bell Atlantic conducted a trial [8] in Pittsburgh called Personal Line Service (PLS). Five hundred users were involved in the test, which was designed to test a single-number, single-phone concept. Handsets and base station units that employed narrowband advanced mobile phone system (NAMPS) technology were used. The trial employed Bell Atlantic's work in advanced intelligent networks (AIN). Bell Atlantic Network Services was to provide the AIN capability that is used to track users, update routing information, and execute service options such as call blocking and call forwarding. Bell Atlantic Mobile provided the cellular capability. Additional call management features that the trial offered were time-of-day routing, personal identification number (PIN), and day-of-week routing. The cost to users participating in this trial included monthly charges for equipment, personal line service, and the AIN function, which included the screening functions, PIN, and blocking functions.

10.6.4 American Personal Communications Trial

American Personal Communications (APC) conducted a trial in the Washington, D.C., and Baltimore areas using CT2 Plus Paging [9] subscriber units. These integrated units were equipped with a numeric pager that enabled the user to receive pages. A button for call-return capability was provided. The cost for users included per-month charges for phone rental and phone service. There was also a per-minute service charge and a per-month paging service charge.

This trial investigated the use of high-mobility (highway speeds), cellular-like services in the 1.8-GHz frequency range. A major goal of the trial was to evaluate interference prediction and avoidance technologies, particularly in environments where coexistence with fixed microwave operations is required. The trial was also designed to demonstrate and measure capacity in a shared RF environment.

APC also tested a low-mobility (pedestrian speeds) service concept in the Washington, D.C., area. Coverage was provided at two airports, in downtown areas, convention centers, and high-use pedestrian areas. Information obtained from this type of service indicated that customers want large, continuous areas of coverage with handoff.

10.6.5 US West Trial

US West advertised plans to conduct a trial with 1,000 people in Boise, Idaho [10–12]. The trial employed home base stations that, in conjunction with handsets, would operate as cordless phones while in the home. Five hundred microcell base stations using frequencies in the 1900-MHz band were planned to be deployed to provide service in public areas. Base stations were to be placed in the city's largest shopping mall, in parks and recreation areas, in residential areas, and along major roads into the downtown area. US West expected the trial to provide a better understanding of a zonal concept that consisted of public, home, and business environments.

10.6.6 Southwestern Bell Personal Communications Inc. Trial

Southwestern Bell announced their intention to conduct a PCN market trial in Houston, employing the intelligent multiple access spectrum sharing (IMASS) system [9,11]. This trial was intended to focus on proving the feasibility of utilizing spectrum gaps in the private operational fixed to 2200 MHz range. Southwestern Bell expected to intelligently assign narrowband PCN channels (less than 5 MHz wide) to those gaps by scanning and identifying microwave channels based on information in a database, then determining the probability of interference from the fixed system based on an interference model, and finally assigning the available channels for use by some set of microcells. Two-way voice capabilities and residential coverage were expected to be tested.

10.6.7 Bell South Trial

Bell South conducted the nation's largest trial, in which 10,000 households were expected to participate [12]. This was also the first trial in which participants were asked to rely exclusively on wireless systems for their home equipment. The main focus of the trial was market and customer needs. In a series of tests using the Bell South cellular network, customers paid for DriveAround (fully mobile, two-way communications), WalkAround (limited mobility, two-way communications), and OutBound (limited mobility, one-way) PCS services. Different types of services were chosen ito measure and understand the demand for different personal communications services and to determine how new services affect existing ones. One interesting result learned from the trial was the extent to which current cellular services and cellular coverage have influenced user expectations. For example, users want fully featured cellular services, roaming, and handoff when they pay cellular prices. Services that differ from the traditional cellular ones are acceptable but only when substantially discounted relative to cellular prices.

10.6.8 Other Trial Activities

The trials described here represent some of the larger, more visible activities underway in the PCS arena. Many more efforts have taken place in the past few years than this short section could address [8]. One indicator of the level of activity is that more than 220 experimental radio licenses have been granted by the FCC for testing personal communications. Additional trials during 1992–1993 were using paging and SMR infrastructures to test PCS concepts. The cable TV industry has also been actively investigating the roles that it could play in supporting and providing PCS in the future. Satellite systems are also under consideration, and access techniques such as CDMA continue to be investigated.

10.6.9 Lessons From the Trials

The trials briefly outlined here are a representative cross section of the many PCS trials that have been conducted. These trials have tested access technologies, market demand, customer service expectations, and willingness to pay. They have tested a variety of service concepts, which differ in their details but which nevertheless embrace the common theme of personal communications services.

In those trials that have tested access technologies, the underlying motivations have been spectral efficiency and sharing the spectrum with other users. The trials that have tested service concepts, however, have focused less on new technology and have made use of existing networks and existing network technologies to realize new services. Since these new services have been trialed in relatively small, controlled environments, the impact on networks caused by their widespread deployment has not yet been gauged. Thus, as service concepts based on user needs become better understood, and as these concepts gain acceptance in the marketplace, investigation of performance characteristics will begin to take place.

Trial experiences have made apparent the user access, communications environment, and service flexibility aspects of the paradigm shifts mentioned in Section 10.1. In one trial [13], for example, the cellular network and fixed network capabilities were made to work together to provide a service that neither of the networks could provide alone. This trial provided useful insights into some of the limitations of those networks and into the types of interworking and service coordination aspects that personal communications may require. This experience suggests the communications network intelligence aspect of the paradigm shift. As our understanding of UPC matures, and as users discover and invent new uses, needs, and applications for their UPC services, and as these uses extend and cross the traditional boundaries of communications networks, it is anticipated that the intelligence and processing power will migrate from centralized locations to the network periphery.

10.7 SUMMARY: KEY CHALLENGES

The personal communications industry will need to address additional challenges before the vision of UPC can be realized. These challenges span the regulatory, technological, business, financial, and sociological areas. In the regulatory area, spectrum licensing must be done via fair methods. Regulatory coordination on a global scale is desirable. In the technological area, advances are needed in device integration and miniaturization; user interfaces; antennas; coding techniques; power storage, management, and consumption; signal processing; and distributed processing. At the system level, complexities associated with distributed intelligence, billing, service creation, service management, and rapid service provisioning will have to be addressed. The business and financial areas face challenges with respect to further identifying and refining the markets; assessing their financial viability; identifying services that will enable exploiting economies of scale; and raising the capital required to build new networks and enhance existing ones. From a social viewpoint, realizing the full potential of UPC for the benefit of the consumer will require careful thought, planning, adaptations, and guarantees of privacy and security. These challenges listed are not the only ones, nor are they isolated and independent of each other.

ACKNOWLEDGMENTS

The authors would like to especially acknowledge the extensive efforts of Richard Zaffino and Julie Brown for their efforts in compiling and contributing source data for this chapter and Jim Chang for his work on the concepts presented here. We would also like to thank Ed Bacher for proofreading early versions and Brian Murphy for source material on Committee T1. We are also grateful for the constructive help provided by Larry Gitten, John Marinho, Joe Nordgaard, Nitin Shah, and several others in the broader AT&T wireless community to ensure correctness of the information included here.

REFERENCES

[1] Brodsky, I., *Personal Communications: The Next Revolution*, Dedham, MA: Artech House, 1994.
[2] Catling, I., *Advanced Technology for Road Transport: IVHS and ATT*, Artech House, 1994.
[3] *PCS News*, April 13, 1993.
[4] Mason, C. F., "GTE Begins New PCS Research Phase," *Telephony*, May 3, 1993. 5. RCR, April 24, 1993.
[5] *RCR*, April 24, 1993.

[6] *Wall Street J.*, May 1994.
[7] Hallman, K., and Czaplewski, "Ameritech's PCS Trial: Integrating Advanced Technologies," *TE&M*, September 1, 1992.
[8] *BIS Strategic Decisions Report*, December 1992.
[9] *PCS Report*, Donaldson, Lufkin & Jenrette.
[10] *RCR*, April 19, 1993.
[11] *Mobile Phone News*, May 17, 1993.
[12] Mason, C. F., "Wireless Fever Rises as SW Bell Unveils Commercial PCS," *Telephony*, April 5, 1993.
[13] *Microwave & RF*, April 1992.

Chapter 11

Personal Communications in Japan and Asia[1]

Megumi Komiya

The Asia-Pacific region can be singled out, in today's global telecommunications market, as the area with the highest growth potential. And given the fact that mobile is one of the few market sectors to escape the current recession, it is hardly surprising that mobile communications in Asia has recently become the focus of much media attention.

Indeed, by the end of this century, the mobile communications market in the Asia-Pacific region is forecast to catch up with, if not surpass, its American and European counterparts in terms of size (Figures 11.1 and 11.2).

Figure 11.3 shows a breakdown of cellular subscribers in various Asian countries. About half of all subscribers are in Japan. However, when one considers the fact that

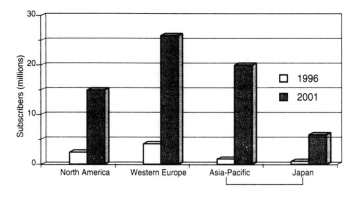

Figure 11.1 Demand forecast for PCN/PCS. (Source: NRI.)

[1]This chapter is reprinted, with minor changes, from *Pan-European Mobile Communications*, Vol. 13, Spring 1993, by kind permission of the author and publisher.

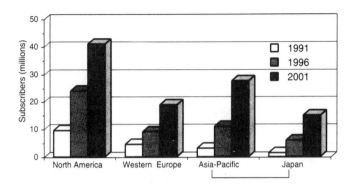

Figure 11.2 Demand forecast for analog/digital cellular. (Source: NRI.)

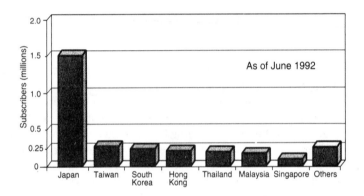

Figure 11.3 Number of cellular subscribers in Asia. (Source: NRI.)

Japan is already a mature cellular market, it seems inevitable that the other Asian countries will enjoy a higher growth rate in years to come.

This chapter focuses first on mobile communications in Japan, exploring the main issues in that market (such as the current market dynamics, the measures introduced to create more of a level playing field, and the switchover from analog to digital.) Then, we will move on to look at the ways in which developments in Japan are likely to affect the mobile scene in other parts of Asia.

11.1 LIBERALIZATION OF THE JAPANESE CELLULAR MARKET

The cellular market in Japan was monopolized by the Nippon Telegraph and Telephone Corporation (NTT) until late 1988, when the Ministry of Posts and Telecommunications (MPT) decided to introduce competition. One other operator was granted permission to enter each of the 10 regional markets in Japan. As shown in Table 11.1, the competition comes from Japan Mobile Communications (JMC) in some areas, while in others it comes from the Daini Denden (DDI) Cellular Group.

Since competition was introduced, NTT has seen an erosion of its market dominance. In the year up to March 1992, NTT's market share was 61.5%, down 1.8% on the previous fiscal year. Over the same period, JMC increased its market share by 1.4%, to 17.7%, and DDI captured a 20.8% share of the market, with an annual growth rate of 0.4%. To put these figures into context, the number of cellular subscribers reached 1.5 million at the end of June 1992.

In July 1992, the Japanese mobile market took another major step away from NTT dominance. The mobile communications division of NTT was separated from its parent company, and a wholly owned subsidiary, NTT DoCoMo, was created. A breakup of this nature had been on the telecommunications ministry's agenda since the privatization of NTT in 1985. But it was only after a long and bitter battle with the MPT that NTT finally agreed to cast off its mobile communications unit. In March 1990, detailed measures were announced to create a level playing field for the newly created company and the non-NTT mobile operators. It should be pointed out that NTT DoCoMo offers not only cellular, but also paging, maritime, and civil aviation communications. As can be seen in Table 11.1, NTT DoCoMo is the only one of the three groups to offer services in all 10 regions of Japan (Figure 11.4 is a map of the regions). This gives the company a tremendous competitive advantage.

However, measures are being introduced to rectify some of these imbalances. First, the MPT has decided that the newly created NTT DoCoMo should be broken up into nine regional companies, among which no cross-subsidization is allowed. These arrangements are due to come into force around July 1993.

Table 11.1

Cellular Network Operators in Japan

Operators	NTT DoCoMo	Japan Mobile Comms	DDI Cellular
Launch date	Dec 1979	Dec 1988	July 1989
Frequencies and standards	800 MHz band NTT	800 MHz band	NTT 800 MHz band J-TACS
	(800 MHz band - digital JDC)	(800 MHz band N-TACS)	(800 MHz band N-TACS)
	(1.5 GHz band - digital JDC)	(1.5 GHz band - digital JDC)	(1.5 GHz band - digital JDC)
Service Area	All regions	Tokyo, Chubu	All regions except Tokyo and Chubu

( ) Have not yet been implemented.
Source: NRI.

Figure 11.4 The 10 regions of Japan. (Source: NRI.)

Second, the networks of all the independent analog cellular operators (JMC and the seven regional subsidiaries of the DDI Cellular Group) will soon be interconnected. Negotiations on financial and technical details are close to completion, and the parties involved are to apply to the telecom ministry for approval. At present, the networks of the eight independent cellular companies are interconnected with NTT's fixed network and NTT DoCoMo's cellular network, but not with each other. Interconnection of the independent networks will surely help to put them more on a par with NTT DoCoMo.

Third, the MPT ruled that the conditions governing interconnection with the NTT fixed network should be exactly the same for both NTT DoCoMo and the independent operators. As a means of eliminating cross-subsidization, the telecom ministry also ruled that NTT DoCoMo must reimburse its parent company in full for the network infrastructure it inherited and for the use of NTT's facilities. However, it is likely to take some time before this materializes.

Fourth, the ministry has decided that the new subsidiary should not enjoy any greater access to NTT's research and development results than what independent operators are allowed. This is significant because NTT has the largest telecommunications R&D resources of any company in modern Japanese history, and this also extends to the mobile sector. Also, NTT is the only company in Japan with the experience of developing a cellular system; other operators have had to either adopt the NTT standard or purchase a system elsewhere (namely, from the United States) and have it modified to suit their needs (i.e., a full turnkey system). The ministry's decision on access to R&D results was intended to prevent the mobile subsidiary from continuing to draw resources from its parent. This ruling has led to the creation of a somewhat unusual feature at NTT DoCoMo: of the 1,800 employees transferred in July 1992 from NTT to its mobile subsidiary, as many as 250 came from the R&D division. It will be interesting to see how this high percentage of R&D staff will affect the new company's commercial performance.

Further competition is to be introduced in the near future. Last year, the standard for a digital cellular system was established, and at the beginning of this year the ministry awarded digital cellular licenses to two more groups of operators: the Digital Phone Group and the Tu-Ka Group (Table 11.2).

NTT DoCoMo will begin offering its digital services in March 1993, and the two new operators will launch theirs in early 1994. In due course, the other current analog operators will also switch over to digital. This means that, by 1995 or 1996, NTT DoCoMo may end up facing competition from three other companies in each of Japan's 10 geographical regions. And all these operators have decided to adopt the same standard: Japanese Digital Cellular (JDC).

11.2 JAPANESE DIGITAL CELLULAR (JDC): A POTENTIAL PAN-ASIAN STANDARD?

Japanese Digital Cellular (JDC) was developed by NTT, in association with various equipment manufacturers. Among those involved were three non-Japanese companies, namely AT&T, Motorola, and Ericsson. Two of these companies—Motorola and Erics-

Table 11.2
New Entrants to the Japanese Cellular Market

Operators	Digital Phone	Tu-Ka Cellular
Launch date	July 1994	Spring 1994
Frequencies and standards	(1.5 GHz band - digital JDC)	(1.5 GHz band - digital JDC)
Service area	Kansai, Tokyo, Tokai	Kansai, Tokyo, Tokai

Source: NRI.

son—recently agreed to cooperate with NTT and the Japanese electronics firm NEC in promoting the JDC standard in Asia. Since these four companies are already collaborating with regard to the Japanese digital cellular network and since they all own complementary technologies, they have decided to jointly pursue JDC business opportunities in Asia. This is a rather intriguing development because Motorola is a proponent of digital advanced mobile phone system (AMPS) in the United States, and Ericsson is deeply involved in the European cellular standard, GSM.

In the summer of 1992, the group organized a series of seminars on JDC in eight Asian countries: Singapore, Thailand, Malaysia, Taiwan, Hong Kong, South Korea, North Korea, and Indonesia. In the seminars, Motorola made a presentation on its voice coding system, Ericsson on its base station facilities, and NEC on its terminal equipment. The technical overview on JDC was provided by an organization called the Research and Development Center for Radio Systems (RCR), which was also in charge of organizing the seminars. The RCR's involvement in promoting JDC stems largely from the fact that it was responsible for setting its technical standard.

What Is the RCR and Why Was It Created?

The RCR is a nongovernmental body, comprising mainly mobile equipment manufacturers and telecommunications carriers. It was established in February 1985 with the aim of developing new wireless systems and setting technical standards.

The RCR grew in importance after the 1986 U.S.-Japan Market-Oriented Sector Selective (MOSS) negotiations on technical standards for wireless facilities. While the setting of all technical standards had previously been the domain of national governments, it was ruled that the governmental role be limited to: (1) ensuring that frequencies are used efficiently, and (2) preventing interference among users. It was also decided that other technical standards should be optional and set by a nongovernmental body, hence the involvement of the RCR. RCR standards are set by a specialized committee in which Japanese and overseas players participate. As of the end of 1991, some 120 companies were participating in the RCR.

Up until 1990, the RCR set standards mainly for those radio systems it had itself developed. However, since 1991, it has also been setting standards for systems in whose development it played no part, namely, the JDC system and second-generation cordless telephone systems.

11.3 MOBILE AS OVERSEAS DEVELOPMENT ASSISTANCE

Yet another interesting development in Japan will affect the development of mobile communications in Asia. In early October 1992, the Japanese telecoms ministry announced that cellular telephone services will be included, for the first time, in its overseas development assistance program. The MPT believes that cellular systems, which can be installed in a short period of time and at a relatively low cost, represent the best way of providing developing countries with a modern communications infrastructure. In fact, many developing countries have already begun installing cellular systems, even though their ordinary telephone networks are often inadequate.

The MPT had intended to complete by the end of 1993 a comprehensive investigation of telecommunications facilities and installation conditions in developing countries, mainly in Asia. It would seek to realize its plans with the help of the Ministry of Foreign Affairs and the Ministry of Finance.

11.4 PROSPECTS FOR DIGITAL CELLULAR IN THE ASIAN MARKET

In the world of commerce, the historical and political relationship between countries is often a good indicator of future commercial trends. Communications is no exception. The classic example of such a relationship is that of color television standards. One can see different spheres of political influence at work in the adoption of the three standards: NTSC, PAL, and SECAM. A similar observation can be made by looking at the different cellular standards adopted by various countries.

In many parts of Asia, the current network infrastructure is composed of elements that are compatible with European technical standards. Decades ago, it was common for Asian countries to import from Europe network infrastructure that had been in use there for some time. Since mobile communications systems need to be connected to the fixed network, the ground may already be prepared in some Asian countries for the introduction of GSM. In fact, the People's Republic of China, Singapore, and India all seem likely to adopt GSM in the near future. And Hong Kong is moving toward a combination of the European and U.S. systems.

As far as cellular is concerned, while the U.S. (AMPS) and European (Nordic mobile telephone, or NMT; GSM; and total access communication system, or TACS) standards have been dominating the world (Figure 11.5), the NTT standard has been adopted by no country other than Japan. Is this situation likely to change, particularly in Asia, with the advent of JDC?

JDC Versus GSM

To explore the answer to that question, one needs to look more closely at the two competing technologies, GSM and JDC. Table 11.3 is a list of technical specifications of the two digital cellular standards.

While GSM was developed to achieve the highest signaling protocols, the engineers who developed JDC were aiming to maximize subscriber capacity, reduce handset size, and offer the system at an economical price.

In terms of frequency utilization, the JDC standard seems to fare better than GSM, and its system capacity also seems to be much larger. In fact, according to one non-

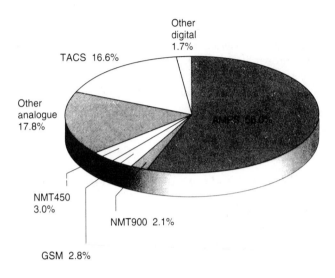

Figure 11.5 Subscriber base by system type. (Source: NRI.)

Table 11.3
Technical Specifications of the Two Digital Cellular Standards

Basic specification comparison		
Frequency band	800 MHz/1.5 GHz	900 MHz
Access method	TDMA	TDMA
Channel spacing	25 kHz	200 kHz
Traffic channels/RF carrier	3 (half rate = 6)	8 (half rate = 16)
Transmission bit rate	42 kbit/s	270.83 kbit/s
Modulation	$\pi/4$ QPSK	GSMK
Voice coding	VSELP	RPE-LTP
General characteristics		
Spectrum	Very highly efficient	Efficient
Utilisation capacity	Very large	Medium
Coverage	Wide area	Wide area

Source: NRI.

Japanese technical specialist, the JDC offers 2.2 times more capacity than GSM and 1.3 times more than digital AMPS.

Another significant difference between the two standards is that JDC has opted for a much lower transmission bit-rate, 42 kbit/s compared with 270.83 kbit/s in the case of GSM.

GSM was designed to focus on the needs of European countries, needs that might be quite different from those of Asian countries. With this in mind, proponents of JDC are trying to argue that their standard better suits the needs of the region because it was developed in an environment similar to that found in many Asian countries (large conurbations with high population densities).

JDC may also receive a boost after the establishment, by July or August 1994, of the half-rate bandwidth voice codec standard. Then, the system capacity of JDC, which already is larger than that of GSM, will double. Supporters of JDC also assert that the cost of handsets will be much lower, though these claims have not yet been substantiated. JDC does, of course, have its technical weaknesses, such as its relatively poor transmission quality in mountainous terrain.

At this time, it is still difficult to predict which digital cellular standard will eventually predominate in Asia. From a purely technical perspective, it may look as though JDC is likely to become the major Asian cellular standard in the future. However, other factors indicate that this might not be the case.

First, many patents regarding the JDC standard are not the sole possession of NTT DoCoMo. In fact, they are owned jointly by NTT DoCoMo and equipment vendors, some of whom are non-Japanese. Nothing has been decided as to how those patents should be handled if the JDC standard is actually adopted by another country. This situation is bound to create complications and disputes. For example, while NTT DoCoMo's partners have agreed to license certain patents free of charge in Japan, the agreements do not extend beyond the country's borders. By contrast, in the case of GSM, these intellectual property rights (IPR) issues have already been sorted out.

Second, all the technical documentation for JDC has been written with only the Japanese market in mind. Many modifications would have to be made before the documentation could be used with ease by other countries seeking to embrace JDC as their standard.

Third, as it stands now, NTT DoCoMo is the only entity in Japan capable of providing support to other countries in the many operational aspects of the JDC system. But the new company's relatively small pool of resources will have to be fully deployed in the domestic market to respond to the fierce competition there. Thus, NTT DoCoMo will not have enough manpower left to provide comprehensive assistance to Asian countries wishing to adopt the JDC standard. The responsibility for this would then fall on the shoulders of JDC equipment manufacturers, who would have to work closely with each other. This might take a while.

11.5 CONCLUSIONS

All things considered, JDC might not fare particularly well in Asia. However, whichever digital standard is adopted, there is no doubt that Japanese mobile equipment vendors possess the technical ability to manufacture suitable terminals. It is likely, therefore, that they will maintain a high share of the digital mobile equipment market in the region.

We will have to keep a very close watch on developments.

Chapter 12
The Future: Third-Generation Mobile Systems

Roger Fudge and John Gardiner

It is a brave person who dares to predict the future direction of any technologically based area of human endeavor. However, the forecaster's task is made somewhat easier for telecommunications in general because such developments require agreements and standards to be set on geographically wide bases, at least internationally. This international agreement takes time, and the timescales are increasing as the necessary breadth of agreement becomes continental and then global. Thus, new systems and services may take about a decade to progress from the initial conception to implementation, and this fact gives us all some warning of what is to come. In this chapter, we survey the current state of progress of the international standards on which the successors to present-day systems will be based.

We use the general term *third-generation mobile systems* (TGMS) to encompass Future Public Land Mobile Telecommunication System (FPLMTS)[1] as a global standard and Universal Mobile Telecommunication System (UMTS) as a European standard. The remainder of this chapter deals with these standards and their relationships, concentrating first on FPLMTS.

12.1 THE NEED

Since the first- and second-generation mobile systems were together expected to satisfy demand for a number of years, it will be advantageous to identify some of the factors that drive the need for a subsequent third-generation system.

There is a trend for telecommunications systems to exhibit a progression from proprietary standards through national standards to international standards. While proprie-

---

[1]At the time of this writing, the ITU was considering proposals to use the term International Mobile Telecommunication-2000 (IMT-2000) instead of FPLMTS.

tary offerings are claimed to be quicker to market, the more widely based standards give manufacturers of proprietary equipment the opportunity to address much larger potential markets. At the same time, the purchasers—both terminal equipment users and the infrastructure network operators—have greater freedom of choice for equipment sourcing. This has a circular effect of reducing prices and increasing performance, which leads in turn to increasing demand. This benefits the end user, who ultimately pays for the service. Thus, more widely accepted standards are likely to drive out more narrowly accepted ones.

There is also a trend toward more complicated systems, although it is not clear whether this is a necessary consequence of development. On one hand, deriving and agreeing on such complex matters on a global basis escalates the magnitude of the task and potentially extends the timescales involved. But this apparent disadvantage of the globalization of standards is balanced by the possibility of involving the worldwide technical community and so marshaling high-quality expertise and increased numbers of people to do the work. Another helpful consequence is that those involved will have an inherent predisposition to adopt and implement the standards they themselves helped to devise; this helps to ensure the large user base that is desirable to gain general acceptance of the new standards.

An additional, and potent, driver is the desire for new services; either as availabilities in more situations and locations (e.g., globally) or in particular for those services that require high information bandwidths. Coupled to this is the desire for smaller, lighter terminals, with longer time between recharging, and lower overall costs.

As mentioned, the timescales involved in producing standards can be long. Figure 12.1 indicates the time it has taken for a number of mobile system standards to progress from concept to implementation and to probable obsolescence.

The advanced mobile phone system (AMPS) was announced in 1971 but was not implemented until 1984. Since it was breaking new ground in many ways, this time lag is perhaps not so surprising. Its close derivative, the total access communication system (TACS), was able to benefit from all the work performed on AMPS and thus had a quicker inception. The pan-European standard GSM also had a long gestation period. Again, this is not too surprising, since work started on GSM before any European cellular systems were in operation. There was also the "path-finding" activity necessary to bring together the manufacturing and operational interests for a number of sovereign states, each with its own language and desires, in such a way that cooperation could be achieved. That it produced *any* standard, let alone one of stature, is a tribute to those who took part in it. It benefited from the foresight of the European regulators, who set aside an allocation of spectrum solely for this system and kept it free of other users. This removed some of the constraints that the designers might otherwise have had to accommodate ito ensure compatibility (or at least coexistence) with other existing systems. This freedom enabled design choices to be optimized, thus reducing timescales.

Thus, TGMS can expect to have a long gestation period. Work started on FPLMTS in 1986, with the first meeting held in Vancouver. The Interim Working Party No. 13

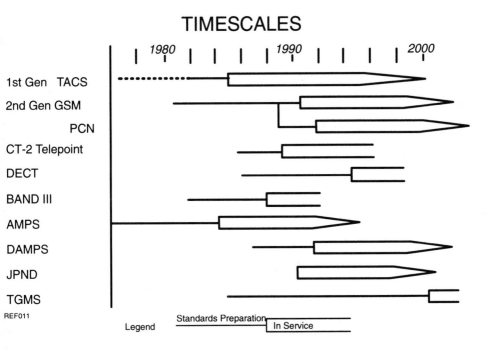

Figure 12.1 Timescales for standards and services.

of Study Group 8 (IWP 8/13) was charged by CCIR (the International Telecommunications Union [ITU] body responsible for radio standards) under its Decision 69 with answering the question, What form should FPLMTS take? The first meeting took place at a time when first-generation cellular systems were in their infancy and many countries had no mobile telephone system. Nonetheless, the likely feasibility of a pocket-sized radio terminal for worldwide use was proposed and largely accepted. It was already apparent that mobile telecommunication systems were evolving rapidly, and that any new generation of equipment could expect to have a prime operational lifetime of some 5 to 10 years.

Thus, with first-generation systems starting in the mid 1980s and second-generation ones scheduled for the early 1990s, it was accepted that TGMS would be required at the turn of the century.

While mobile radio telephony, certainly in its cellular form, was in its infancy in 1986, there were strong indications of its growth potential. But telephony, in the sense of being connected to the fixed-telephone networks, was far from being the only form of mobile radio communications. There was, therefore, the question of incorporating within FPLMTS the needs of other mobile radio users, for example, emergency services (police, fire, ambulance) and dispatch services, such as taxi fleets and trucking companies. While these requirements are not included in the usual understanding of ''public'' nor do they necessarily need to interface with the public switched telephone networks (PSTNs), there could well be benefits from their inclusion in FPLMTS. These benefits would be manifest to the user as a lower cost service compared to a dedicated one and to the regulator as a better use of frequency spectrum. Therefore, such requirements were included in the services that FPLMTS should support, with the understanding that the merits of doing this would be apparent only when detailed system design factors and system costings were available much later in the story of FPLMTS.

It was recognized that careful consideration would have to be given to the range of services encompassed by FPLMTS. On the one hand there was the desire to incorporate as wide a range as possible; on the other, there was the need to temper this desire with a strong degree of realism.

One factor of the user's needs that was highlighted early in the evolution of FPLMTS was that of quality, in all its aspects:

- Fidelity of transmission;
- Ease of access to services, including availability in the sense of time (e.g., lack of call blocking) and geography (wide coverage);
- Suitability of hardware;
- Ease of use.

Many of these factors are easy to state, in broad terms, and are easy to agree on in committees and working parties. It is much more difficult to translate them into agreed-on hard numerical specifications, but that is the necessary function of the standards bodies involved in the task.

The need for radio spectrum to be available for the operation of FPLMTS was also recognized from the beginning. At first, this was unquantified in terms of amount, band-width, or location in the spectrum. Frequency availability is always a sensitive matter; it is allocated by administration (states) within guiding principles agreed on through the ITU, which is an organ of the United Nations. Any change in use of the spectrum, to accommodate a new major service type, has to recognize the time it takes to make such changes and the impact on the current users of any segment of spectrum that will be redesignated. The facts that FPLMTS was being studied by an organ of the ITU and that it was expected to involve worldwide use were helpful in the early stages of obtaining suitable spectrum. The ITU had the foresight to put in its calendar of events a World Administrative Radio Conference for 1992 (WARC '92) to look at the spectrum allocation in the range 1 GHz to 3 GHz. This range could accommodate the likely extensive bandwidth desires of FPLMTS and would also be acceptable to operators from considerations of propagation and implementation. It was necessary to tread carefully in the process of requesting WARC '92 to allocate spectrum to FPLMTS, because the issue was not an isolated one but had to be considered in relation to the needs of all spectrum users and to take into account many other matters that were to be considered at the WARC. Thus, the desires of the engineers involved in FPLMTS to make an early and substantial bid to WARC '92 for spectrum had to be tempered by the needs of the administration to be diplomatic in all respects. Nevertheless, WARC '92 presented the only chance of obtaining suitable frequency allocation—on a worldwide basis—so the chance could not be lost!

12.2 THE START

As already mentioned in Section 12.1, work on FPLMTS for CCIR started in 1986, and it quickly identified the personal station (PS)—a personal pocket-sized radio—as a key feature. But such a small device would surely be subject to inherent limitations on its range from a base station and on its operating duration, due to its limited battery capacity. Thus, small cells would be necessary to support its operation—the so-called microcell—and there could be problems for the network in supporting the rapid handover function necessary to allow a PS to travel at high speeds, as would be desired by users in a car, for example. The concept arose of arranging for the vehicle (car, bus, truck, ship, or aircraft) to have its own base station (BS) to feed the PS; this BS would in turn be connected to the (fixed) mobile network infrastructure via another radio interface specifically identified for mobile vehicle use. In these situations, then, the PS user would make use of two radio connections in tandem.

The cells associated with vehicular use would be considerably larger than those for the PS, and network economics would support the provisioning of such cells in wider geographical areas than for the microcellular environment. But even so, service to areas of low user density would still not be economically viable as part of the initial view of FPLMTS, based on the simple mobile user–terrestrial base station concept.

At this point, the desires of two other interests in communications for the future made their voices clear. The satellite system operators (notably the International Maritime Satellite Organization (INMARSAT)) wanted to have their interests supported by FPLMTS, and the developing countries foresaw that delivering telecommunications to their populations, whether of low density in rural areas or high density in urban areas, could benefit greatly from the ease and low cost of developing radio access rather than installing wires. Thus, they saw FPLMTS as providing an economical alternative to deploying a fixed wired network, and they were conscious of their populations' need for telecommunications.

Thus, the concept of FPLMTS grew from its simple cellular origins to a relatively complex structure that would encompass a satellite mode of delivery and provide access to telecommunications services for fixed users. These latter two factors may well come to be seen as highly important aspects of FPLMTS rather than subsidiary roles. The satellite mode will provide not only worldwide coverage, but it also will help "grow" the usage of FPLMTS on the edges of terrestrial coverage to the point where terrestrial coverage becomes economic, on a rolling basis. The fixed use of FPLMTS is being increasingly regarded as an economical means of providing the local loop connection for fixed usage and also as being the way of satisfying the emerging enormous demands for telecommunications connection for the fixed user in developing countries. The satellite component will have a role to play in this fixed service, largely because of its ability to provide virtually instantaneous service to an installed terminal (see also Chapter 9).

Growth in other directions also occurred. Since the core terminal was a small lightweight device, with the implication that it would have very low cost by virtue of a common design and very large user base worldwide, it would be sensible to use it in a domestic or business cordless mode. Moreover, a large portion of the private mobile radio (PMR) service requirements could be supported by the microcell or larger cellular structure.

12.3 FIXED AND MOBILE CONVERGENCE: UNIVERSAL PERSONAL TELECOMMUNICATION

The mobility element of FPLMTS had been the reason for CCIR starting work on the subject, since mobility was regarded as needing, or at least implying, a radio link to the user. It became apparent during the work on FPLMTS that mobility could itself be classified as a service. But this raised the question of the definition of mobility for this situation. While CCIR regarded mobile communications as being within its domain, a sister body under ITU—CCITT—saw itself as particularly responsible for fixed telecommunications. Thus, CCIR thought of fixed telecommunications networks as sources and sinks for the calls made over FPLMTS and as a means for supporting the mobility given by FPLMTS. CCITT, on the other hand, viewed mobile networks as tails to the fixed network, allowing "its" telecommunications users the freedom of mobility. Both, of course were right, the difference being one of standpoint.

After considerable discussion, it was agreed that there were two forms of mobility. Terminal mobility was characterized by the concepts of FPLMTS that provided for mobility of the (radio) terminal. User mobility viewed users as being mobile and able to access telecommunication services from any terminal, fixed or mobile, and to obtain from that terminal a set of services configured to their particular requirements. Thus, one form of mobility allows a user with a radio terminal to move freely and continuously within a spatial area or volume, while the second provides users, who need to have some form of identification and authenticity, with mobility in a discrete sense anywhere worldwide where they can access a suitable terminal.

The latter concept has adopted the mantle of *personal communications* since it refers to the person, not the terminal, and to the fact that the services to be made available at that terminal, both incoming and outgoing, are specifically configured to the needs of the person. Thus, the terminal acquires a "personality." This concept is defined by CCITT as universal personal telecommunications (UPT), in which a central concept is that of a user having a personal telephone number that identifies him or her for incoming and outgoing calls made to or from any terminal in any location and that also identifies the user's profile of services and billing arrangements. CCIR has agreed that FPLMTS will support UPT.

Those working on FPLMTS were conscious that, when the system is implemented around the year 2000, in most areas the novelty of using a telephone while mobile, as occurred in first- and second-generation systems, would no longer seem novel. Users would regard such mobile telephones as just "plain old telephones" that were not restrained by wires. They would expect services and quality equivalent to those offered on tethered telephones. Thus, to users there would be just a telephone service—and naturally they could take their telephone instruments with them whenever desired. There would be no distinction from a user perspective between a fixed network and a mobile network.

In addition, the engineers designing the mobile and fixed network architecture saw increasing overlap of their interests and a blurring of the ability to define an interface between what was a mobile network and one that was fixed. Such a distinction was becoming more of a commercial and regulatory matter. The commonality of function in the network increased dramatically when the implications of implementing UPT were realized. Such functionality required the ability to locate and authenticate a mobile user— just the functions that mobile networks had traditionally regarded as their special requirements. There is, therefore, a merging of interests of the fixed and the mobile network operators, with the likely convergence of hardware and software.

12.4 THE FORM OF FPLMTS

The FPLMTS concept is described pictorially in Figure 12.2, which illustrates the ability of FPLMTS to operate flexibly in diverse environments, including that depending on service provision via satellite.

FPLMTS is being defined by a series of CCIR recommendations (these will become Recommendations of the Radiocommunications Sector of the ITU following its reorganization) that are in various stages of approval and that will evolve and expand with time.

The recommendations are categorized into a series of layers going from the general to the detailed and specific. The titles given to the layers are: concept, requirements, framework, selection procedures, selection, and specification. Such a staged process is considered necessary to accommodate the complexity of the subject and the attendant operational, regulatory, and manufacturing considerations. The process also recognizes that the ITU, even in its new structure, is not a body that can by itself simply issue standards and ensure their compliance worldwide.

The topics that are considered to merit separate Recommendations across the layers are: services (i.e., the telecommunication services to be supported by FPLMTS), radio interfaces (the topic of greatest concern to the ITU-R), satellite issues, the needs of the developing countries, network architecture and network interfaces, network management, and security procedures.

The work by the CCIR, initially by IWP8/13 and later by Task Group 1 of Study Group 8 (TG8/1), took the CCIR, and to some extent the ITU, into new territory, which required amending the traditional form of its work and the structure of its reports and recommendations. Some of these are further explored here since they provide elements for the description of FPLMTS, which is itself complex and extensive.

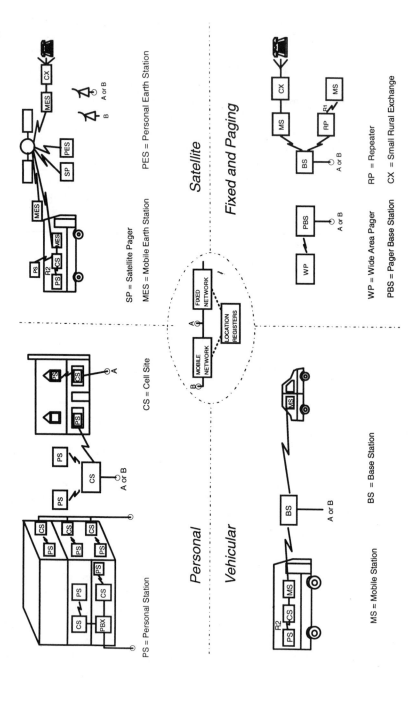

Figure 12.2 FPLMTS in different environments.

A brief description of FPLMTS is taken from Section 1 of ITU Recommendation 816 on Services for FPLMTS [1]:

Future Public Land Mobile Telecommunications Systems (FPLMTS) are third-generation mobile systems (TGMS) which are scheduled to start service around the year 2000. They will provide access, by means of one or more radio links, to a wide range of telecommunications services supported by the fixed telecommunication networks (e.g. PSTN/ISDN), and to other services which are specific to mobile users.

A range of mobile terminal types is encompassed, linking to terrestrial or satellite based networks, and the terminals may be designed for mobile or fixed use.

Key features of FPLMTS are:

- incorporation of a variety of systems
- high degree of commonality of design worldwide
- compatibility of services within FPLMTS and with the fixed networks
- use of a small pocket terminal worldwide.

It was agreed at an early stage in CCIR that the work should ideally be led by the user and the market, rather than derived from "technical push"—a desire much easier to state than to achieve.

Traditionally, CCIR had dealt with radio matters, and these had been handled by the national administrations within their own countries. CCITT had dealt largely with the telegraph and telephone interests of the same administrations, largely contained within national monopoly telecommunications operators (post, telephone, and telegraph authorities, or PTTs). But now many countries were allowing competition in their fixed networks and allowing provision of services over these networks by third parties. Even more countries regulated for competition in the provision of mobile telecommunications, and the move to the so-called deregulated environment, for both fixed and mobile operation and services, was growing. Therefore, FPLMTS had to recognize the needs of the multiple operators, the split between network operator and service supplier, regulatory and deregulatory influences, varying pace of change of different administrations, the cascading of networks (terrestrial to satellite to fixed, etc.), and, above all, the growing dominance of the end user in determining requirements. New ground, therefore, has been broken within the ITU and, consequently, for all current and future participants.

The telecommunications services provided by FPLMTS are numerous and varied, including the means to carry both digital and digitized analog (e.g., sound and video) signals. These in turn support derived services such as paging, messaging, access to voicemailbox, facsimile, multimedia, and videophony. In the first phase of FPLMTS, user bit rates up to approximately 2 Mbit/s will be accommodated, and later phases will extend the services and admit higher bit rates. There is a strong desire to make available to the user as many integrated services digital network (ISDN) services as possible. This would ideally be via a transparent B channel (32 kbit/s), but it is not clear whether this is economical in cost or spectrum terms. If it were achievable, then it would meet the firm requirement for high quality; an objective could be to make it equivalent to the quality of 32 kbit/s adaptive differential pulse code modulation (ADPCM) operating in the fixed network.

The functional architecture of FPLMTS is based on that of the intelligent network (IN), as in Figures 12.3 and 12.4, which are derived from ITU Recommendation 817 [2].

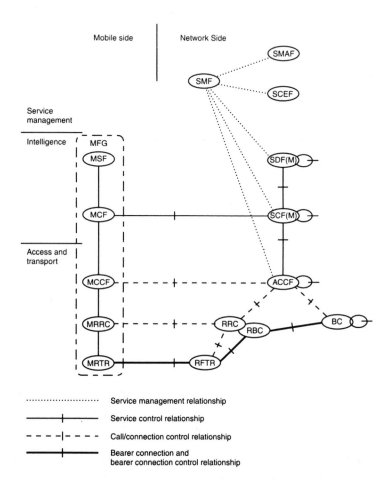

Figure 12.3 Functional model for FPLMTS. (After ITU Rec. 817.) ACCF, access and call control function; BC, bearer control; MCCF, mobile call control function; MCF, mobile control function; MFG, mobile functions group; MRRC, mobile radio resource control; MRTR, mobile radio transmission and reception; MSF, mobile storage function; RBC, radio bearer control; RFTR, radio frequency transmission and reception; RRC, radio resource control; SCEF, service creation environment function; SCF(M), service control function (mobile); SDF(M), service data function (mobile); SMAF, service management access function; SMF, service management function.

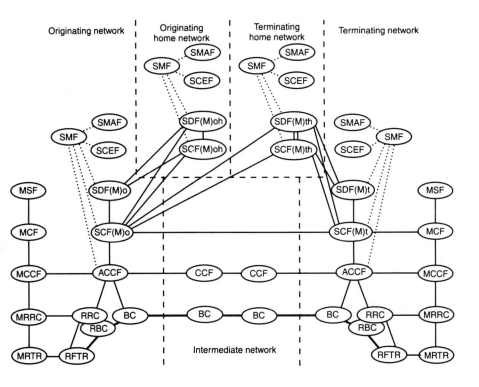

Figure 12.4 Network functional reference model. (After ITU Rec. 817.) See Figure 12.3 for abbreviations.

It was noted in Section 12.1 that WARC '92 presented a major opportunity for spectrum allocation to be made available to FPLMTS; in fact, WARC '92 specifically asked how much was needed and where it should be located. The IWP 8/13 working on FPLMTS debated this matter extensively, although some administrations were concerned about the authority of the group to make a decision about this question. Most administrations agreed, however, that IWP8/13 was the competent body to make the estimation, at a technical level. To assess the requirements, it was necessary to assume traffic levels, grade of service, and the bit rate of speech codecs. For the critical urban operational areas, these were taken as 1,500 Erlangs/km^2, 1% blocking probability, and 16 kbit/s, respectively, although it was recognized at the time that such starting assumptions were far from being firm. This was particularly troublesome for the bit rate of the speech codec, where the assumptions regarding the speech quality requirement, including allowable speech delay, have a considerable impact on the bit rate.

From the assessment, it was concluded that FPLMTS needed 230 MHz as a minimum for the terrestrial segment, but many service requirements were not included in this estimate, so that it is not really even a lower bound. Of this 230 MHz, 60 MHz was associated with the PS, and this is the minimum that was considered to be needed as a worldwide allocation to support PS. IWP 8/13 was cautious in requesting a specific location for this spectrum, since this was deemed to be the province of the regulatory authorities. To give them maximum flexibility, an allocation in the range 1 to 3 GHz was suggested.

WARC '92 decided that the bands 1185 to 2015 MHz and 2100 to 2200 MHz should be "identified for use by FPLMTS by those administrations wishing to implement the systems." This was not on an exclusive basis, and sharing with other mobile services and the incumbent fixed services in these bands could be necessary. Many countries and administrations around the world envisage using these bands, in full or in part, for FPLMTS. The United States has, however, already assigned a major portion of these bands to their burgeoning mobile telecommunications industry for personal communications services (PCS) use. The portion of the bands from 1980 to 2010 MHz and 2170 to 2200 MHz were designated by WARC '92 for mobile satellite services and by implication to the satellite component of FPLMTS.

12.5 RELATIONSHIP WITH THE EUROPEAN ACTS RESEARCH PROGRAM

Reference was made in Chapter 1 to the research programs undertaken on a pan-European basis and funded via the European Commission. It is relevant to return to this subject in the context of TGMS since the RACE programs particularly, while focusing initially on background research for second-generation systems, have now moved forward to more advanced research to underpin the definition of TGMS standards. At the time of writing, a major successor program to RACE is in the final stages of preparation under the 4th Framework Program for Europe and is concentrating on exploring TGMS services and applications in an advanced communications technologies and services program (ACTS).

The emphasis by ACTS on services and applications raises an interesting and fundamental issue in relation to how TGMS will interface with second-generation systems. Will TGMS follow an essentially evolutionary path, or will the third generation be a completely novel and revolutionary communications environment? To answer this question requires first a definition of revolution. Frequent reference was made in earlier chapters to the application of new technologies in future air-interfaces, particularly the attractions of CDMA, but deployment of CDMA rather than TDMA in the physical layer hardly constitutes a revolution, given that the radio elements of second-generation systems are by no means the major area of investment and functionality in what are now complex system architectures. It must also be borne in mind that the commitments made to second-generation systems by the operators in more than 60 countries worldwide are bound to have an influence on the ways in which TGMS technology will be introduced and supported.

The work of the ETSI Special Mobile Group (SMG) subgroups has already been alluded to in the context of TGMS, but there is also a significant level of activity in SMG5 (the subgroup responsible for work on third-generation systems) in promoting evolution through enhanced versions of second-generation systems, in particular GSM. Many aspects of the UMTS standard as they have been defined thus far will probably be accommodated in "GSM-plus," suggesting that the further extrapolation of UMTS to FPLMTS might be an essentially evolutionary process. At the same time, the view taken by the European Commission is that UMTS must possess unique and distinctive features. The apparently different positions taken by the Commission on the one hand and the operators on the other are reconcilable, however, if the distinctive capabilities of UMTS/TGMS are recognized as "access to bandwidth." A "revolution" is then necessary if service transparency across the radio–fixed network boundary is to be achieved as user expectations continue to expand to embrace multimedia services of all sorts, but for the operators, an evolutionary approach toward service transparency at the ISDN basic rate level still represents major advance. Indeed, given the level of investment in second-generation systems, coexistence of second- and third-generation systems will be inevitable with evolved second-generation systems supporting parts of the UMTS environment and truly revolutionary elements of TGMS supporting services with bandwidth demands from ISDN primary rate (2.048 Mbit/sec) upward.

On the basis of this philosophy, the Commission's emphasis on trials of services and applications in ACTS becomes increasingly understandable, given that much of the experimental program will be exploring new types of service supported by second-generation technology. The ACTS program, therefore, takes as its starting point the three platforms of UMTS; the mobile broadband service (MBS), perceived as occupying spectrum at 60 GHz; and the emerging high-bit-rate wireless LAN standard Hiperlan. These three platforms owe their existence to the second phase of the RACE program. The ACTS work plan reflects this approach in the way in which the possible range of activity is divided into seven areas, covering the principal topics of interest listed in Chapter 1, Section 1.4.2 with the addition of horizontal actions, which relates to such aspects as the concertation process, technology training programs, and promoting the exploitation of results.

In budgetary terms, it is expected that ACTS will be supported by resources several times greater than those allocated to RACE. In bringing the second-generation systems to fruition in Europe, the member states took a decisive lead in the mobile and personal communications arena. It is clearly the Commission's intention that this position shall be maintained in the global markets for TGMS products in the early years of next century.

12.6 CONCLUSIONS

Third-generation mobile telecommunications will be seen by the using public as just a telephone system, one that is not tethered to a fixed point by a wire and that has a wide range of services, wide availability, and low cost. For the satellite component of FPLMTS, at least, these services will be available literally worldwide. The quality and range of services will in general be closely comparable to those provided by the then contemporary fixed network. FPLMTS have the potential to provide a viable telecommunication system for the mobile and fixed users in developing countries.

Network operators and manufacturers of network equipment will find a high degree of commonality between the mobile and fixed systems in terms of hardware and software, which will be manifest to users in the form of a low network usage cost.

All bodies involved in the work on FPLMTS profess a desire to cooperate toward a common end, and the achievement of this goal in an environment of commercial and national pressures will be a fitting tribute to the efforts of those who continue to refine the concepts and the standards which embody them.

ACKNOWLEDGMENTS

Extracts from ITU publications [1,2] are reproduced with the prior authorization of ITU as copyright holder. However, their selection is the sole responsibility of the author. The complete volumes of the ITU material can be obtained from: International Telecommunication Union, General Secretariat—Sales and Marketing Service, Place des Nations, CH-1211 Geneva 20, Switzerland.

REFERENCES

[1] ITU Rec 816, "Framework for Services Supported on Future Public Land Mobile Telecommunication Systems (FPLMTS)," RM 1992.
[2] ITU Rec 817, "Future Public Land Mobile Telecommunication Systems (FPLMTS)," RM 1992.

Glossary

This glossary collects abbreviations and selected technical terms used in the book, plus some others commonly encountered.

ACTS   Advanced communications technologies and services: a European research initiative under the Fourth Framework Program.

ADPCM Adaptive differential pulse code modulation: a method of digitally encoding speech signals.

ALOHA A simple multiple access protocol invented at the University of Hawaii in which users transmit whenever they have something to send. A variant that offers greater throughput is "slotted" ALOHA, in which transmissions are synchronized to a universal clock.

AMPS   Advanced mobile phone system: the American analog cellular telephone system.

ANSI   American National Standards Institute.

ATM   Asynchronous transfer mode: a packetized digital transfer system, adopted for the B-ISDN. (Beware: the same abbreviation is used for automated teller machines, i.e., "hole-in-the-wall" bank cash machines.)

BER   Bit error rate.

BHCA Busy hour call attempts.

B-ISDN Broadband integrated services digital network.

BPSK Binary phase shift keying.

CCIR   Comité Consultatif International de Radio: formerly the ITU body responsible for radio standards (now the responsibility of ITU-R).

CCITT Comité Consultatif International Télégraphique et Téléphonique: formerly the ITU body responsible for nonradio standards (now the responsibility of ITU-T).

CDF	Cumulative distribution function: the integral of the PDF.
CDMA	Code-division multiple access.
CDCS	Continuous dynamic channel selection: a channel management technique used in DECT.
CEN	Comité Européen de Normalisation (European Committee for Standardization).
CENELEC	Comité Européen de Normalisation Electrotechnique (European Committee for Electrotechnical Standardization).
CEPT	Conférence Européenne des Administrations des Postes et des Télécommunications: the European ''club'' of PTTs, which, before ETSI was created, was a focus of technical standards setting in Europe.
CERS	Communications Engineering Research Satellite: a U.K. satellite project.
C/I	Carrier-to-interference ratio, usually expressed in dB.
COST	Cooperation in Science and Technology: a European forum for the exchange of research results in various areas. The COST-231 project relates to the evolution of mobile radiocommunication systems, including propagation aspects.
CSMA	Carrier sense multiple access: a multiple-access protocol that offers improved performance over ALOHA, users being required to listen for a quiet channel before transmitting.
CT0, CT1	Early cordless telephone standards.
CT2	Second-generation cordless telephone.
CT3	Third-generation cordless telephone.
CTR	Common technical regulations: the basis for type-approval of, for example, GSM handsets.
CUG	Closed user group.
CW	Carrier wave, that is, a constant, unmodulated radio carrier.
DAMPS	Digital AMPS: a digital cellular system having some limited compatibility with the (analog) AMPS system (U.S.).
dB	Decibel, a logarithmic measure of power ratio.
dBi	dB relative to isotropic: a measure of antenna gain.
dc	Direct current, that is, zero frequency.
DCS1800	Digital communication system: a variant of the GSM standard providing for operation in the 1800-MHz band, initially required by the United Kingdom for its PCN service.
DECT	Digital European Cordless Telecommunications: the second-generation cordless system standardized by ETSI.
DFE	Decision feedback equalizer.

DLC	Data link control (layer).
DRX	Discontinuous reception.
DS	Direct sequence: a form of spread-spectrum system, using a pseudo-random binary stream to spread the signal (see Chapter 8).
DSRR	Digital short range radio: a second-generation radio communication system for applications not requiring infrastructure support.
DTI	Department of Trade and Industry: U.K. government body responsible for regulating telecommunications.
DTMF	Dual-tone multi-frequency: system of low-speed signaling in telephone systems, for example, for dialing, using paired audio tones.
DTX	Discontinuous transmission.
ERC	European Radiocommunications Committee.
Erlang	Unit of telecommunication traffic intensity, equal to an average of one call in progress.
ERMES	European radio message system: second-generation paging system standardized by ETSI.
ERO	European Radiocommunications Office.
ETS	European Telecommunications Standard.
ETSI	European Telecommunications Standards Institute.
FCC	Federal Communications Commission: U.S. government regulatory body.
FDDI	Fiber distributed data interface: a U.S. standard for high-rate fiber-optic token-ring LAN systems.
FDMA	Frequency-division multiple access.
FEC	Forward error correction.
FH	Frequency hopping: a form of spread-spectrum system (see Chapter 8).
FPLMTS	Future public land mobile telecommunication system: the ITU name for third-generation systems. Since this name is neither memorable nor pronounceable in any language, the name IMT-2000 has been proposed.
GEO	Geostationary earth (satellite) orbit (see Chapter 9).
GHz	Gigahertz.
GMSK	Gaussian minimum shift keying: a form of constant-envelope binary digital modulation.
GoS	Grade of service: in telephony, the probability that a call will not succeed. Note that a high GoS is worse than a low one. Also used loosely to mean service quality.
GSM	Groupe Spécial Mobile: originally, the CEPT (later, ETSI) committee responsible for the pan-European digital cellular standard. Also, used as the name of the system and the service. Global System for

Mobile (Communication): the name for the GSM system and service, invented by a group of European operators, to fit the abbreviation "GSM."

GWSSUS — Gaussian wide sense stationary, uncorrelated scattering.

HEO — Highly elliptical (satellite) orbit (see Chapter 9).

IBCN — Integrated broadband communications network.

IEEE — Institution of Electrical and Electronic Engineers (U.S.). IEEE-802 is a committee responsible for developing standards for LANs.

IMT-2000 — International mobile telecommunications-2000: a name proposed within ITU for their third-generation concept, otherwise known as FPLMTS.

IN — Intelligent network.

IPR — Intellectual property rights, including patents, trademarks, copyrights.

IR — Infrared.

ISDN — Integrated services digital network.

ISI — Intersymbol interface.

ISM — Industrial, scientific, and medical (bands): spectrum allocations for those services.

ISO — International Standards Organization.

ITU — International Telecommunication Union.

IVHS — Intelligent vehicle highway system.

JDC — Japanese Digital Cellular (see Chapter 11).

kHz — Kilohertz.

LAN — Local area network.

LED — Light-emitting diode.

LEO — Low earth (satellite) orbit (see Chapter 9).

LOS — Line of sight (radio path).

LPI — Low probability of intercept: a property of spread-spectrum systems.

MAC — Medium access control (protocol layer).

MAN — Metropolitan area network.

MAP — Mobile application part: an extension of signaling system number 7, providing support of mobile systems.

MCN — Microcellular network.

MEO — Medium earth (satellite) orbit (see Chapter 9).

MoU — Memorandum of Understanding: often in reference to a group of European GSM operators who in 1987 signed an MoU providing for cooperation on commercial and operational matters of common interest and who are known as the "GSM/MoU Group."

MHz — Megahertz.

NET — Norme Européenne de Télécommunications: formerly, the specification for type-approval, produced by CEPT, later replaced by CTRs.

	Thus, NET-10 was the type-approval spec for GSM, now replaced by CTR-5 and CTR-9.
NMT	Nordic mobile telephone (system): cellular telephone system prevalent in the Nordic countries.
NPRM	Notice of Proposed Rule Making: an official pronouncement by the FCC (U.S.).
NTT	Nippon Telegraph and Telephone Corporation (Japan).
OFTEL	Office of Telecommunications: U.K. official body created to protect the interests of consumers of telecommunication services.
OKQPSK	Offset-keyed quadrature phase-shift keying: digital modulation system in which in-phase and quadrature components carry bit-streams offset by half a bit, resulting in desirable spectral and envelope characteristics.
OSI	Open systems interconnection: the ISO layered protocol model.
PABX	Private automatic branch exchange.
PAD	Packet assembler/disassembler: device used to interface with a packet network.
PBX	Private branch exchange.
PCIA	Personal Communications Industry Association (U.S.).
PCN	Personal communications network: used as a general term for such networks, and specifically for the U.K. systems licensed for operation in the 1800-MHz band and using the DCS1800 standards.
PCS	Personal communication service (or system).
PDA	Personal digital assistant: a name coined to describe a portable, screen-based communication-oriented organiser.
PDF	Probability distribution function.
PDH	Plesiochronous digital hierarchy: transmission system standard (plesiochronous = near-synchronous).
PHP	Personal handy phone: a Japanese standard for cordless telephony.
PIN	Personal identification number: a (typically four-digit) secret number to be input by the user to obtain service.
PLMN	Public Land Mobile Network.
PMR	Private mobile radio.
POTS	Plain old telephone service.
PSK	Phase-shift keying.
PSPDN	Packet-switched public data network.
PSTN	Public-switched telephone network.
PTO	Public telecommunications operator.
PTT	Post, telephone and telegraph (authorities): an old name for the (usually state-monopoly) operators of telecommunication and postal services.

QAM	Quadrature amplitude modulation.
QoS	Quality of service: an ill-defined term covering various measures of "quality" in telecommunication systems.
QPSK	Quadrature phase-shift keying.
QWSS	Quasi-wide-sense stationary.
RACE	R&D in Advanced Communications Technologies in Europe.
RCR	R&D Center for Radio Systems (Japan) (see Chapter 11).
RELP	Residually excited linear prediction: the speech encoding method used in the GSM full-rate coder.
RES	Radio equipment and systems: a technical committee of ETSI, responsible for terrestrial radio standards other than GSM.
RMS	Root-mean-square (value): of a statistical distribution, the same as standard deviation.
SDH	Synchronous digital hierarchy: a transmission system standard.
SFH	Slow frequency hopping.
SMG	Special Mobile Group: the name adopted by ETSI for the (former) GSM committee, for consistency with other ETSI Technical Committees, which all had English names. There are numerous subgroups: SMG1, SMG2, etc. The SMG5 subgroup has responsibility for work on third-generation systems.
SNR	Signal-to-noise ratio, usually expressed in dB.
S-PCN	Satellite personal communications system.
SS7	Signaling system number 7: an ITU standard for telecommunications network signalling.
T1	U.S. standards committee, active in personal communications area (see Chapter 10).
TACS	Total access communication system: an analog cellular phone system based on AMPS, used in the United Kingdom and elsewhere.
TBR	Technical basis for regulation.
TCM	Trellis coded modulation.
TDD	Time-division duplex: two-way communication using synchronized alternate transmission on a single carrier.
TDMA	Time-division multiple access.
TETRA	Trans-European trunked radio: second-generation digital PMR system standardized by ETSI.
TFTS	Terrestrial flight telephone service: digital telephone service for aircraft passengers standardized by ETSI.
TGMS	Third-generation mobile systems.
TIA	Telecommunications Industry Association (U.S.).
TRX	Transmitter/receiver.
T-SAT	Technology Satellite: a U.K. satellite project.

UHF	Ultra high frequency: usually defined as 0.3 to 3 GHz.
UMTS	Universal mobile telecommunication system (or service): the concept of third-generation systems developed in Europe, particularly by the RACE program.
UPT	Universal personal telecommunication.
VA	Voice activation.
WAN	Wide area network.
WARC	World Administrative Radio Conference.
WLLAN	Wireless local area network.
WSS	Wide-sense stationary.
WSSUS	Wide-sense stationary, uncorrelated scattering.

GSM has generated its own language of abbreviations and acronyms, of which the following are only a selection. Some of these terms have come to be used outside the context of the GSM standards and systems.

A interface	The interface between the BSC and the MSC. A-bis interface (originally called the B interface) is the internal interface within the base station (BSS) between the BTS and the BSC.
A3	The authentication algorithm.
A5	The ciphering algorithm.
A8	The cipher key computation algorithm.
AoC	Advice of Charge (supplementary service).
AGCH	Access grant channel.
AuC	Authentication center.
BAIC	Barring of All Incoming Calls (supplementary service).
BAOC	Barring of all Outgoing Calls (supplementary service).
BCCH	Broadcast Control Channel.
BIC-roam	Barring of Incoming Calls when roaming outside the home country (supplementary service).
BOIC	Barring of Outgoing International Calls (supplementary service).
BOIC-exHC	Same as BOIC except for calls to home country.
BSC	Base station controller.
BSIC	Base station identity code.
BSS	Base station subsystem (i.e., BSC plus one or more BTSs).
BTS	Base transceiver station.
CBCH	Cell broadcast channel.
CFB	Call Forwarding on mobile subscriber Busy (supplementary service).
CBSMS	Cell-broadcast short message service.
CFB	Call Forwarding on mobile subscriber Busy (supplementary service).
CFNRc	Call Forwarding on mobile subscriber Not Reachable (supplementary service).

CFNRy	Call Forwarding on No Reply (supplementary service).
CFU	Call Forwarding Unconditional (supplementary service).
CLIP	Calling Line Identification Presentation (supplementary service).
CLIR	Calling Line Identification Restriction (supplementary service).
CoLP	Connected Line Identification Presentation (supplementary service).
CoLR	Connected Line Identification Restriction (supplementary service).
CW	Notification of Call Waiting (supplementary service).
EIR	Equipment identity register.
FACCH	Fast-associated control channel.
FCCH	Frequency correction channel.
HLR	Home location register.
IMEI	International mobile equipment identity.
IMSI	International mobile subscriber identity.
MAP	Mobile application part.
MCC	Mobile country code.
MPTy	Multiparty (supplementary service).
MS	Mobile station.
MSC	Mobile services switching center.
NCC	Network color code.
NSS	Network and switching subsystem (not an official GSM term).
OSS	Operation subsystem.
PAGCH	Paging and access grant channel.
RACH	Random access channel.
SACCH	Slow associated control channel.
SCH	Synchronization channel.
SIM	Subscriber identity module (''smart card'').
SMS	Short-message service.
TCH	Traffic channel.
TCH/F	Traffic channel/full-rate (for 13 kbit/s speech or 3.6/6/12 kbit/s data).
TCH/H	Traffic channel/half-rate (for 7 kbit/s speech for 3.6/6 kbit/s data).
TEI	Terminal equipment identity.
TMSI	Temporary mobile subscriber identity.
TRAU	Transcoder/rate adapter unit.
VLR	Visitor location register.

About the Authors

The Editors

John Gardiner graduated with first-class honors from Birmingham University (UK) and subsequently obtained a PhD in 1964. After working for several Racal companies, he joined Bradford University in 1968 as a lecturer. He is currently head of the Department of Electronic and Electrical Engineering and leader of the Mobile and Personal Communications Research Group.

Professor Gardiner is a consultant in radiocommunications to the Telecommunications Division of the UK Department of Trade and Industry (DTI). Between 1989 and 1991 (and also during early 1994) he was a visiting expert at DGXIII of the European Commission in Brussels, where he has been active in the formation of Community policy towards introduction of new technology in the mobile and personal communications sector.

A Fellow of the IEE and the Royal Academy of Engineering, he is currently a member of Professional Group E8. He has contributed to four books and has published 180 papers on research topics in radio communications.

Barry West is a freelance consultant operating in the areas of mobile radio and personal communication systems. A current responsibility is the coordination of the UK Government-sponsored LINK Personal Communications Programme, which promotes joint industrial/academic research projects in the personal communications area.

Mr. West was formerly head of R&D and Standards for Mercury PCN Ltd. (now Mercury One-2-One); before that he was with the GEC-Marconi Research Centre, where he led the Mobile Communications Group. He holds the degree of MSc in telecommunication systems from Essex University.

The Authors

Stephen Barton spent the early part of his career in industrial research with Marconi Research and Signal Processors Ltd., and in UK Government service with the Government

Communications Centre and the Rutherford Appleton Laboratories. From 1989 to 1994 he was Professorial Fellow in signal processing at the University of Bradford, and was appointed professor of signal processing in February 1994.

His main research interests are in signal processing with application to personal communications, wireless local area networks and satellite systems.

Professor Barton has a BSc(Eng) degree in electrical engineering from University College London (1970) and an MSc in Telecommunication Systems from the University of Essex (1974). He is a Fellow of the IEE and a senior member of the IEEE.

Roger Fudge started his career with the Post Office Research Station, and later moved to Marconi Space and Defence Systems to design satellite systems and components. His involvement with mobile radio began at the UK Home Office, working on radio communication systems for the police and fire services. He joined British Telecom Mobile Communications as chief engineer in 1984, working on the design of cellular systems including TACS and GSM. Since 1986, he has been active in the work of standards bodies in this area, particularly ITU Task Group 8/1. Currently, he continues to be involved in this topic, from a satellite perspective, on behalf of INMARSAT.

Dr. Fudge holds the degree of BSc(Eng) from the University of London and a PhD from City University, London. He is a Fellow of the IEE.

Trevor Gill is chief engineer with Vodafone Ltd., one of the two UK operators of cellular radio systems. He is responsible for a group developing computer tools for propagation prediction, frequency planning and network optimization.

He joined Racal Research in 1977 to work on VHF and HF radio systems, digital speech, and adaptive antennas. Involvement with mobile telephone systems began in 1983, with research into the choice of technology for the UK analog networks to be built for Cellnet and Vodafone. Subsequently, Mr. Gill worked on systems design for CT2 equipment and contributed to the creation of standards for the GSM system. He transferred to Vodafone in 1991.

Mr. Gill holds the degree of MA in electrical sciences from Cambridge University (UK), and is a member of the IEE.

Hans van der Hoek graduated in economics from Rotterdam University in 1986, and went on to help set up and run the business school at Delft University of Technology.

In 1990, Mr. van der Hoek joined Ericsson Business Mobile Networks BV in Amsterdam, The Netherlands, as product manager for cordless telephone systems, and in 1991 he was appointed Global Manager, Distribution and Sales for Ericsson's cordless telephone systems and business mobile networks business.

Megumi Komiya is a director of strategic research at Nomura Research Institute Europe in London. Dr. Komiya received her PhD in telecommunications policy from Michigan State University and taught international telecommunications there. She was also an associate at Kalba International, a Boston-based international telecommunications consulting firm before she moved to London. Dr. Komiya has published articles on a range of subjects including the comparative analysis of ISDN developments in Japan and the US, strategic management of telecommunications, and industrial policy analysis of the Japanese computer industry.

Anil Kripalani is a department head at AT&T Bell Laboratories, Whippany, New Jersey, responsible for AT&T's wireless systems and services architecture, and also their integrated access architecture. Since joining Bell Laboratories in 1976, he has been involved with the design and development of packet switching networks, value added networks, and since 1984, cellular and personal communications network infrastructure equipment for the AMPS and GSM markets. He holds a number of patents in this area.

He received his Bachelor's degree in electrical engineering from the Indian Institute of Technology, New Delhi, India in 1974, and a Master's degree in computer science from the University of California in 1975. Mr. Kripalani is chairman of the Mobile and Personal Communications Standards 1800 Committee (TR-46) of the Telecommunications Industry Association (TIA). He is a member of the IEEE.

Kaveh Pahlavan is the Waren Hadden Professor of Electrical and Computer Engineering and the director of the Center for Wireless Information Network Studies at the Worcester Polytechnic Institute (WPI), Worcester, Massachusetts. His recent research has been on indoor radio propagation modeling, and analysis of multiple access and transmission methods for wireless local networks.

He started his career as an assistant professor at Northeastern University, Boston, and before joining WPI, he was the director of advanced development at Infinite Inc., Andover, Massachusetts, working on voice band data communications.

Professor Pahlavan is the editor-in-chief of the *International Journal on Wireless Information Networks*. He was the program chairman and organizer of the 1991 IEEE Wireless LAN Workshop and the 1992 IEEE International Symposium on Personal, Indoor, and Mobile Radio Communications. He is a member of Eta Kappa Nu and a senior member of the IEEE Communication Society.

Robin Potter is engineering director at Mercury One-2-One, one of the two licensed PCN operators in the UK. He has responsibility for the network design and planning and has been involved with the development of the DCS1800 PCN standard since the start of this company. Prior to his work on PCN, Mr. Potter was involved with the development of the DCS1800 PCN standard headed research in Mobile Systems at British Telecom, after many years' involvement in the development of digital switching systems.

Peter Ramsdale is head of technical strategy at Mercury One-2-One, one of the two licensed PCN operators in the UK. He has responsibility for the standards used in PCN and the technical basis for the development of a personal communications business. Dr.

Ramsdale has been involved in PCN since its inception, having previously managed the radio department of STC Technology Ltd. (STL), which was responsible for considerable innovation in radio design and implementation.

Jesse Russell is AT&T's chief wireless architect and director of the AT&T Bell Laboratories Wireless Systems and Services Architecture Center.

Since joining Bell Labs in 1972, Mr. Russell's responsibilities have spanned cellular communication systems, digital remote switching, traffic control, loop transmission, and network management. He holds several key patents in the area of wireless communications.

Mr. Russell received a BSEE degree from Tennessee State University in 1972 and an MS from Stanford University. He is an IEEE Fellow and has served on the board of governors of the IEEE Vehicular Technology Society. He currently chairs the Mobile and Personal Communications Division of the Telecommunications Industry Association (TIA), and is president of the Electromagnetic Energy Association. He is a member of Eta Kappa Nu and Tau Beta Pi honor societies.

Stephen Temple started his career as an apprentice with the Marconi Company (UK), subsequently working on satellite ground station systems for Marconi's Space Division. He entered UK government service in 1971, eventually becoming deputy director of engineering in the Home Office Directorate of Telecommunications. Since 1984 he has been director of technical affairs in the Telecommunications Division of the UK Department of Trade and Industry (DTI), where he is responsible for telecommunications, radio and broadcasting matters.

He was closely involved with the conception of the GSM cellular radio system, and worked with others to reform European telecommunications standards-making. In 1988 he was elected chairman of the Technical Assembly of ETSI, the newly-formed European Telecommunications Standards Institute.

Mr. Temple has a BSc in electronics from Southampton University and is a Fellow of the IEE. In 1994 he received the IEEE Award in International Communication for his contributions to worldwide telecommunications standards.

Adel Turkmani was assistant professor of the American University of Beirut, Lebanon from 1986–87, and in 1987 he moved to the University of Liverpool (UK), where he became a senior lecturer. In 1994 he joined Mobile Systems International, London.

Dr. Turkmani is the author of 60 technical papers in the area of indoor and outdoor radio propagation, channel characterization, man-made noise and diversity techniques.

He holds the degree of PhD from the University of Liverpool, and is a member of the Institution of Electrical Engineers (UK), and a member of IEEE.

Index

The Artech House Telecommunications Library

Vinton G. Cerf, Series Editor

Smart Cards, José Manuel Otón and José Luis Zoreda

Super-High-Definition Images: Beyond HDTV, Naoisha Ohta, Sadayasu Ono, and Tomonori Aoyama

Television Technology: Fundamentals and Future Prospects, A. Michael Noll

Telecommunications Technology Handbook, Daniel Minoli

Telecommuting, Osman Eldib and Daniel Minoli

Telemetry Systems Design, Frank Carden

Telephone Company and Cable Television Competition, Stuart N. Brotman

Teletraffic Technologies in ATM Networks, Hiroshi Saito

Terrestrial Digital Microwave Communications, Ferdo Ivanek, editor

Toll-Free Services: A Complete Guide to Design, Implementation, and Management, Robert A. Gable

Transmission Networking: SONET and the SDH, Mike Sexton and Andy Reid

Transmission Performance of Evolving Telecommunications Networks, John Gruber and Godfrey Williams

Troposcatter Radio Links, G. Roda

Understanding Emerging Network Services, Pricing, and Regulation, Leo A. Wrobel and Eddie M. Pope

UNIX Internetworking, Uday O. Pabrai

Virtual Networks: A Buyer's Guide, Daniel D. Briere

Voice Processing, Second Edition, Walt Tetschner

Voice Teletraffic System Engineering, James R. Boucher

Wireless Access and the Local Telephone Network, George Calhoun

Wireless Data Networking, Nathan J. Muller

Wireless LAN Systems, A. Santamaría and F. J. López-Hernández

Wireless: The Revolution in Personal Telecommunications, Ira Brodsky

Writing Disaster Recovery Plans for Telecommunications Networks and LANs, Leo A. Wrobel

X Window System User's Guide, Uday O. Pabrai

For further information on these and other Artech House titles, contact:

Artech House
685 Canton Street
Norwood, MA 02062
617-769-9750
Fax: 617-769-6334
Telex: 951-659
email: artech@world.std.com

Artech House
Portland House, Stag Place
London SW1E 5XA England
+44 (0) 171-973-8077
Fax: +44 (0) 171-630-0166
Telex: 951-659
email: bookco@artech.demon.co.uk